Introductory Mathematics for Earth Scientists

By the same author and published by Dunedin Academic Press:
Mathematical Modelling for Earth Sciences (2008)
ISBN: 9781903765920

INTRODUCTORY MATHEMATICS FOR EARTH SCIENTISTS

Xin-She Yang

Department of Engineering, University of Cambridge

DUNEDIN

Published by
Dunedin Academic Press ltd
Hudson House
8 Albany Street
Edinburgh EH1 3QB
Scotland

www.dunedinacademicpress.co.uk

ISBN 978-1-906716-0-04

First published 2009, reprinted 2011

BRITISH LIBRARY CATALOGUING IN PUBLICATION DATA
A catalogue record for this book is available from the British Library

*While all reasonable attempts have been made to ensure the accuracy of information
contained in this publication it is intended for prudent and careful professional and
student use and no liability will be accepted by the author or publishers for any loss,
damage or injury caused by any errors or omissions herein. This disclaimer does not
effect any statutory rights.*

Printed and bound by CPI Group (UK) Ltd, Croydon, CR0 4YY

Contents

Preface

Any quantitative analysis in earth sciences requires mathematical analysis to a certain degree, and many mathematical methods are essential to the modelling and analysis of the geological, geophysical and environmental processes widely studied in earth sciences. In response to such needs in modern earth sciences, the author wrote a book on *Mathematical Modelling for Earth Sciences* published by Dunedin Academic in 2008 with the intention of filling the gaps between mathematical modelling and its practical applications in earth sciences. The responses of enthusiastic readers and professional reviews of the book in scientific journals such as *Geoscientist* and *Journal of Sedimentary Research* are very encouraging and positive. We realised that *Mathematical Modelling for Earth Sciences* is most suitable for the third and final year undergraduates with more mathematical training, graduates and researchers. As we understand from various surveys and interactions with earth scientists, the mathematical training of new undergraduates often begins at the secondary school or GCSE level. This may pose substantial challenge in teaching a wide range of mathematical skills along with the specialised subjects in earth sciences. This present book is an attempt to bridge the gaps between secondary mathematics and mathematical skills at university level.

This introductory book strives to provide a comprehensive introduction to the fundamental mathematics that all earth scientists need. The book is self-contained and provides an essential toolkit of basic mathematics for earth scientists assuming no more than a standard secondary school level as its starting point.

The topics of earth sciences are vast and multidisciplinary, and consequently the mathematical tools required by students are diverse and complex. Thus, we have to strike a fine balance between coverage and detail. Topics have been selected to provide a concise but comprehensive introductory coverage of all the major and popular mathematical methods. The book offers a 'theorem-free' approach with an emphasis on practicality. With more than a hundred step-by-step worked examples and dozens of well-selected illustrative case studies, the book is specially suitable for non-mathematicians and geoscientists. The topics include binomial theorem, index notations, polynomials, sequences and series, trigonometry, spherical trigonometry, complex numbers, vectors and matrices, ordinary differential equations, partial differential equations, Fourier transforms, numerical methods and geostatistics.

Mathematics is an exact science; however, the ability and skills to be able to do estimations are also crucially important in many applications in earth sciences. Many processes in earth sciences are coupled together and it is difficult, even impossible in most cases, to get the exact solutions or answers to a problem of interest. In this case, some estimations using simple formulae, rather than complex mathematics, often provide more insight to the underlying mechanism and processes. To get the right magnitudes or orders of some quantities by simple estimation is an art that needs development by practice. For example, we know the tectonic plate drifts at a speed of about a few centimetres per year, this can be obtained by using the age difference and distances among the Hawaii Islands. The magnitude is a few centimetres per year, not millimetres per year or kilometres per year. If we want a higher accuracy, we have to use complex partial differential equations, the actual geometry of the plate, the detailed knowledge of the Earth's interior, the mantle convection mechanism, and many other inputs. We also have to make sure that we have the right mathematical models and the appropriate solution techniques both analytically and numerically. In the end, we may produce seemingly more accurate solutions. However, we may still not believe the results as there are so many factors by which they can be affected. In this case, a simple estimate is very useful and provides a good basis to build more complex or realistic models. In this book, we will provide dozens of examples to guide you in using basic mathematics to actually model important processes and evaluate the right orders or magnitudes of solutions. The case studies include the Airy isostasy, the flexural deflection of a tectonic plate, greenhouse effect, Brownian motion, sedimentation and Stokes' flow, free-air and Bouguer gravity, heat conduction, the cooling of the lithosphere, earthquakes and many others.

Introductory Mathematics for Earth Scientists introduces a wide range of fundamental and yet widely-used mathematical methods. It is designed for use by undergraduates, though postgraduate students will also find it a helpful reference and aide-memoire.

In writing the book, I sent more than 100 questionnaires to many university professors, lecturers and students at various universities including Oxford, Cambridge, MIT, Imperial, Princeton, Caltech, Colorado, Cornell, Leeds, and Newcastle (Australia). Their valuable comments to my questionnaires concerning the essential mathematical skills of earth scientists, selection of mathematical topics, and their relevance to applications have been incorporated in the book. In addition, I have received a variety of help from my colleagues and students. Special thanks to my students at Cambridge University: Joanne Perry, Edward Flower, Alisdair McClymont, Omar Kadhim, Mi Liu, Maxine Jordan, Priyanka De Souza, Michelle Stewart, Shu Yang, Young Z. L. Yang, and Qi Zhang.

Last but not least, I thank my wife and son for their help.

Xin-She Yang
Cambridge 2009

Chapter 1

Preliminary Mathematics I

The basic requirement for this book is a good understanding of secondary mathematics, about the GCSE[1] level mathematics. All the fundamental mathematics will be gradually introduced in a self-contained style with plenty of worked examples to aid the understanding of essential concepts.

We are concerned mainly with the introduction to important mathematical techniques; however, we have to deal with physical quantities with units. We will mainly use SI units throughout this book. For example, we use kg/m^3 for density, metrer (m) for length, seconds (s) for time, m/s for velocity, m/s^2 for acceleration, Pascals (Pa) for pressure and stress, Pascal-seconds (Pa s) for viscosity, and Kelvin (K) or degree Celsius (°C) for temperature. The unit of geological time is often expressed in terms of million years (Ma), and occasionally, years (yr or a). As for the latter, yr becomes a popular usage, though a was a formal usage. Therefore, we will use yr for simplicity.

The applications of mathematics in earth sciences are as diverse as the subject itself. For a particular problem, we often have to use many different mathematical techniques, and more often their combinations, to obtain a good solution to the problem of interest. For example, in order to obtain the deflection of the lithosphere under a given load as discussed in later chapters, we have to use almost all the important techniques introduced in this book, including basic mathematics, trigonometry, complex numbers, and differential equations.

On the other hand, a good mathematical technique can be applied to a wide range of problems. For example, simple manipulations of

---

[1] General Certificate of Secondary Education, achieved usually at age 16 in the UK by a vast majority of students in any subject before proceeding to A level studies and/or undergraduate studies.

1

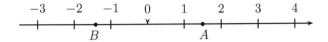

Figure 1.1: Real numbers and their representations (as points or locations) on the number line.

index forms and basic algebraic equations can be applied in studying the greenhouse effect, glacier flow, isostasy and the estimation of the size of an impacting crater, as will be demonstrated later in this chapter.

1.1   Functions

1.1.1   Real Numbers

From basic mathematics, we know that whole numbers such as -5, 0 and 2 are integers, while the positive integers such as $1, 2, 3, \ldots$ are often called natural numbers. The ratio of any two non-zero integers p and q in general is a fraction p/q where $q \neq 0$. For example, $\frac{1}{2}$, $-\frac{4}{3}$ and $\frac{22}{7}$ are fractions. If $p < q$, the fraction is called a proper fraction. All the integers and fractions make up the rational numbers.

Numbers such as $\sqrt{2}$ and π are called irrational numbers because they cannot be expressed in terms of a fraction. Rational numbers and irrational numbers together make up the real numbers. In mathematics, all the real numbers are often denoted by $\Re$, and any real number corresponds to a unique point or location in the number line (see Fig. 1.1). For example, $3/2$ corresponds to point A and $-\sqrt{2}$ corresponds to point B.

In a plane, any point can be located by a pair of two real numbers (x, y), which are called Cartesian coordinates (see Fig. 1.2). For ease of reference to a location or different part on the plane, we conventionally divide the coordinate system into four quadrants. From the first quadrant with $(x > 0, y > 0)$ in an anti-clockwise manner, we consecutively call them the second, third and fourth quadrants as shown in Fig. 1.2. Such representation makes it straightforward to calculate the distance d, often called Cartesian distance, between any two points $A(x_i, y_i)$ and $B(x_j, y_j)$. The distance is the line segment AB, which can be obtained by the Pythagoras's theorem for the right-angled triangle ABC. We have

$$d = \sqrt{(x_i - x_j)^2 + (y_i - y_j)^2} \geq 0. \qquad (1.1)$$

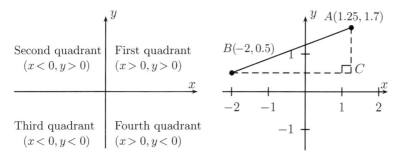

Figure 1.2: Cartesian coordinates and the four quadrants.

Example 1.1: For example, the point $A(1.25, 1.7)$ lies in the first quadrant, while point $B(-2, 0.5)$ lies in the second quadrant. The distance between A and B is the length along the straight line connecting A and B, and we have

$$d = \sqrt{[1.25 - (-2)]^2 + (1.7 - 0.5)^2} = \sqrt{12.0025} \approx 3.4645,$$

which should have the same unit as the coordinates themselves. Here the sign '$\approx$' means 'is approximately equal to'.

1.1.2 Functions

A function is a quantity (say y) which varies with another independent quantity x in a deterministic way.

For example, the permeability of a porous material is closely related to its grain size by a quadratic function

$$K = CD^2, \tag{1.2}$$

which is the Hazen's empirical relationship between permeability and grain size D. In generally, the larger the grain size, the larger the pores or voids, and thus the higher the permeability. Let $D \approx D_{10}$ be Hazen's effective size or diameter (in cm). Then, $C \approx 100$ if K is in the unit of cm^2, while $C \approx 0.01$ if K has a unit of m^2. For example, for very fine sand and fine sand, $D \approx 0.0625$ to 0.5 mm (or 1/16 to 1/2 mm). Using $D = 0.1$ mm $= 0.01$ cm, we have

$$K \approx 0.01 \times (0.01)^2 = 10^{-6} \text{m}^2. \tag{1.3}$$

Furthermore, the volume y of a sphere with a radius x is simply

$$y = \frac{4\pi}{3} x^3, \tag{1.4}$$

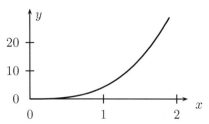

Figure 1.3: The graph of the function $y = f(x) = 4\pi x^3/3$.

which is an example of a cubic function. For any given value of x, there is a unique corresponding value of y. By varying x smoothly, we can vary y in such a manner that the point (x, y) will trace out a curve on the $x - y$ plane (see Fig. 1.3). Thus, x is called the independent variable, and y is called the dependent variable or function. Sometimes, in order to emphasize the relationship, we use $f(x)$ to express a general function, showing that it is a function of x. This can also be written as $y = f(x)$.

Example 1.2: The average density of the Earth can be calculated by $\rho = M_\odot/V$, where $M_\odot \approx 5.979 \times 10^{24}$ kg is the mass of the Earth, and V is the volume of the Earth. Since the radius of the Earth is about $R = 6.378 \times 10^3$ km $= 6.378 \times 10^6$ m, the volume is

$$V = \frac{4\pi}{3}R^3 \approx 1.087 \times 10^{21} \text{ m}^3,$$

so the mean density is approximately

$$\rho = \frac{5.979 \times 10^{24}}{1.087 \times 10^{21}} \approx 5.502 \times 10^3 \text{ kg/m}^3 = 5.502 \text{ g/cm}^3.$$

This is much higher than the density (typically 2.6 g/cm³) of rocks near the Earth's surface, which implies that the density of the Earth's interior should be much higher, hence may be a very dense core.

For any real number x, there is a corresponding unique value $y = f(x) = 4\pi x^3/3$, and although the negative volume is meaningless physically, it is valid mathematically. This relationship is a one-to-one mapping from x to y. In mathematics, when we say x is a real number, we often write $x \in \Re$. Here $\Re$ means the set of all real numbers, and $\in$ stands for 'is a member of' or 'belongs to'. Thus, $x \in \Re$ is equivalent to saying that x is a real number or a member of the set of real numbers.

The domain of a function is the set of numbers x for which the function $f(x)$ is defined validly. If a function is defined over a range $a \le x \le b$, we say its domain is $[a, b]$ which is called a closed interval.

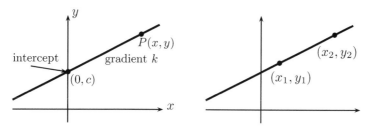

Figure 1.4: The equation $y = kx + c$ and a line through two points.

If a and b are not included, we have $a < x < b$ which is denoted as (a, b), and we call this interval an open interval. If b is included while a is not, we have $a < x \le b$, and we often write this half-open and half-closed interval as $(a, b]$. Thus, the domain of function $f(x) = x^2$ is the set of all real numbers $\Re$ which is also written

$$-\infty < x < +\infty. \tag{1.5}$$

Here ∞ denotes infinity. The values that a function can take for a given domain is called the range of the function. Thus, the range of $y = f(x) = x^2$ is $0 \le y < \infty$ or $[0, \infty)$.

Sometimes, it is more convenient to write a function in a concise mathematical language, and we write

$$f : x \mapsto \frac{1}{2}x^2 - 15, \qquad x \in \Re, \tag{1.6}$$

which is called a mapping. This means that $f(x)$ is a function of x and this function will turn any real number x (input) in the domain into another number (output) $\frac{1}{2}x^2 - 15$.

The simplest general function is probably the linear function

$$y = kx + c, \tag{1.7}$$

where k and c are real constants. For any given values of k and c, this will usually lead to a straight line if we plot values y versus x on a Cartesian coordinate system. In this case, k is the gradient of the line and c is the intercept (see Fig. 1.4).

For any two different points $A(x_1, y_1)$ and $B(x_2, y_2)$, we can always draw a straight line. Its gradient can be calculated by

$$k = \frac{y_2 - y_1}{x_2 - x_1}, \qquad (x_1 \ne x_2). \tag{1.8}$$

For any point $P(x, y)$, the equation for the straight line becomes

$$\frac{y - y_1}{x - x_1} = k = \frac{y_2 - y_1}{x_2 - x_1}. \tag{1.9}$$

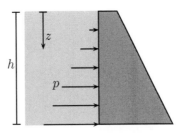

Figure 1.5: Pressure variations along the wall of a dam.

This is

$$y = y_1 + \frac{(y_2 - y_1)}{(x_2 - x_1)}(x - x_1) = \frac{(y_2 - y_1)}{(x_2 - x_1)}x + [y_1 - \frac{(y_2 - y_1)}{(x_2 - x_1)}x_1], \quad (1.10)$$

which gives the intercept

$$c = y_1 - \frac{(y_2 - y_1)}{(x_2 - x_1)}x_1. \quad (1.11)$$

In a special case $y_1 = y_2$ while $x_1 \neq x_2$, we have $k = 0$. The line becomes horizontal, and the equation simply becomes $y = y_1$ which does not depend on x as x does not appear in the equation.

In another special case, $x_1 = x_2$, k is very large or approaching infinity, which we write as $k \to \infty$. It becomes a vertical line. Any small change in x will lead to an infinite change in y. For example, if we have $x_1 = x_2 = 2$, we have a vertical line through $x = 2$, and we simply write the equation as $x = 2$, and y does not appear in the equation in this case.

Example 1.3:   The water pressure p in a dam varies with depth z

$$p(z) = \rho g z,$$

where ρ is the density of water, and g is the acceleration due to gravity. The linear increase of pressure requires the dam thickness to increase with depth so as to balance the pressure by shear strength of the dam (see Fig. 1.5). As the pressure is a linear function of z, the average pressure $\bar{p}$ can be calculated by

$$\bar{p} = \frac{p(0) + p(h)}{2} = \frac{0 + \rho g h}{2} = \frac{\rho g h}{2}.$$

A special function is the modulus function $|x|$ which is defined by

$$|x| = \begin{cases} x & \text{if } x \geq 0, \\ -x & \text{if } x < 0. \end{cases} \quad (1.12)$$

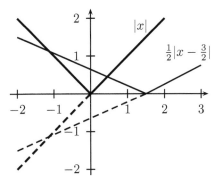

Figure 1.6: Graphs of modulus functions.

That is to say, $|x|$ is always non-negative. For example, $|5| = 5$ and $|-5| = 5$.

The modulus function has the following properties

$$|a \times b| = |a| \times |b|, \qquad a, b \in \Re, \tag{1.13}$$

which means that $|a^2| = |a|^2 = a^2 \geq 0$ for any $a \in \Re$. Similarly, for any real numbers a and $b \neq 0$, we have

$$\left|\frac{a}{b}\right| = \frac{|a|}{|b|}. \tag{1.14}$$

But for addition in general, we have $|a+b| \neq |a| + |b|$. In fact, there is an equality in this case: $|a+b| \leq |a| + |b|$, and we will discuss this in detail in the vector analysis in later chapters.

Two examples of modulus functions $f(x) = |x|$ and $f(x) = \frac{1}{2}|x - \frac{3}{2}|$ are shown as solid lines in Fig. 1.6. The dashed lines correspond to x and $\frac{1}{2}(x - \frac{3}{2})$ (without the modulus operator), respectively.

As an example, we now try to calculate the time taken for an object to fall freely from a height of h to the ground. If we assume the object is released with a zero velocity at h, we know from the basic physics that the distance h it travels is

$$h = \frac{1}{2}gt^2, \tag{1.15}$$

where t is the time taken, and $g = 9.8$ m/s^2 is the acceleration due to gravity. Now we have

$$t = \sqrt{\frac{2h}{g}}. \tag{1.16}$$

The velocity just before it hits the ground is $v = gt$ or

$$v = \sqrt{2gh}. \tag{1.17}$$

For example, the time taken for an object to fall from $h = 20$ m to the ground is about $t = \sqrt{2 \times 20/9.8} \approx 2$ seconds. The velocity it approaches the ground is $v \approx \sqrt{2 \times 9.8 \times 20} \approx 19.8$ m/s.

Example 1.4: The escape velocity of a satellite can be calculated as follows: The kinetic energy of a moving object is

$$E_k = \frac{1}{2}mv^2,$$

where m is the mass of the satelite, and v is its velocity. The potential energy due to the Earth's gravitational force is

$$V = -\frac{GM_E m}{r},$$

where M_E is the mass of the Earth, r is the radius of the Earth, and G is the universal gravitational constant.

The total energy must be zero if the object is just able to escape the Earth. We have

$$E_k + V = \frac{1}{2}mv^2 - \frac{GM_E m}{r} = 0.$$

This means that

$$v = \sqrt{\frac{2GM_E}{r}}.$$

Since the gravity g on the Earth surface is

$$g = \frac{GM_e}{r^2},$$

we have

$$v = \sqrt{2gr}.$$

Using the values of $g = 9.8$ m/s^2, and $r = 6370$ km$= 6.4 \times 10^6$ m, we have

$$v = \sqrt{2 \times 9.8 \times 6.4 \times 10^6} \approx 11170 \text{ m/s} = 11.17 \text{ km/s}.$$

Interestingly, the escape velocity is independent of the mass of the object.

Quadratic functions are widely used in many applications. In general, a quadratic function $y = f(x)$ can be written as

$$f(x) = ax^2 + bx + c, \tag{1.18}$$

where the coefficients a, b and c are real numbers. Here we can assume $a \neq 0$. If $a = 0$, it reduces to the case of linear functions that we have

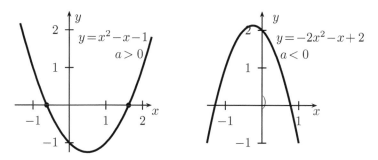

Figure 1.7: Graphs of quadratic functions.

discussed earlier. Two examples of quadratic functions are shown in Fig. 1.7 where both $a > 0$ and $a < 0$ are shown, respectively.

Depending on the combination of the values of a, b, and c, the curve may cross the x-axis twice, once (just touch), and not at all. The points at which the curve crosses the x-axis are the roots of

$$f(x) = ax^2 + bx + c = 0. \tag{1.19}$$

As we can assume $a \neq 0$, we have

$$x^2 + \frac{b}{a}x + \frac{c}{a} = 0. \tag{1.20}$$

Of course, we can use factorisation to find the solution in many cases. For example, we can factorise $x^2 - 5x + 6 = 0$ as

$$x^2 - 5x + 6 = (x - 2)(x - 3) = 0, \tag{1.21}$$

and the solution is thus either $(x - 2) = 0$ or $(x - 3) = 0$. That is $x = 2$ or $x = 3$. However, sometimes it is difficult (or even impossible) to factorise even seemingly simple expressions such as $x^2 - x - 1$. In this case, it might be better to address the quadratic equation in a generic manner using the so-called complete-square method.

Completing the square of the expression can be used to solve this equation, and we have

$$x^2 + \frac{b}{a}x + \frac{c}{a} = (x + \frac{b}{2a})^2 - \frac{b^2}{4a^2} + \frac{c}{a}$$

$$= (x + \frac{b}{2a})^2 - \frac{(b^2 - 4ac)}{4a^2} = 0, \tag{1.22}$$

which leads to

$$(x + \frac{b}{2a})^2 = \frac{(b^2 - 4ac)}{4a^2}. \tag{1.23}$$

Taking the square root of the above equation, we have either

$$x + \frac{b}{2a} = +\sqrt{\frac{b^2 - 4ac}{4a^2}}, \quad \text{or} \quad x + \frac{b}{2a} = -\sqrt{\frac{b^2 - 4ac}{4a^2}}. \quad (1.24)$$

Using $\sqrt{4a^2} = +2a$ or $-2a$, we have

$$x = -\frac{b}{2a} \pm \frac{\sqrt{b^2 - 4ac}}{2a} = \frac{-b \pm \sqrt{b^2 - 4ac}}{2a}. \quad (1.25)$$

In some books, the expression $b^2 - 4ac$ is often denoted by Δ. That is $\Delta = b^2 - 4ac$. If $\Delta = b^2 - 4ac < 0$, there is no solution which corresponds to the case where the curve does not cross the x-axis at all. In the special case of $b^2 - 4ac = 0$, we have a single solution $x = -b/(2a)$, which corresponds to the case where the curve just touches the x-axis. In general, if $b^2 - 4ac > 0$, we have two distinct real roots. Let us look at an example.

Example 1.5:   For the quadratic function $f(x) = x^2 - x - 1$, we have $a = 1$, $b = -1$ and $c = -1$. Since

$$\Delta = b^2 - 4ac = (-1)^2 - 4 \times 1 \times (-1) = 5 > 0,$$

it has two different real roots. We have

$$x_{1,2} = \frac{-(-1) \pm \sqrt{\Delta}}{2} = \frac{1 \pm \sqrt{5}}{2}.$$

So the two roots are $(1 - \sqrt{5})/2$ and $(1 + \sqrt{5})/2$ which are marked as solid circles on the graph in Fig. 1.7.

Of course, many expressions involve higher-order terms such as $x^5 - 2x^3 - 4x^2 + 1$. In this case, we are dealing with polynomials which use the notation of indices and manipulations of equations extensively.

1.2   Equations

When we discussed the quadratic function $f(x)$ earlier, we have to find where it crosses the x-axis. In this case, we have to solve the quadratic equation $ax^2 + bx + c = 0$. In general, an equation is a mathematical statement that is written in terms of symbols and an equal sign '='. All the terms on the left of '=' are collectively called the left-hand side (LHS), while those on the right of '=' are called the right-hand side (RHS). For example, $x^2 + 2x - 3 = 0$ is a quadratic equation with its left-hand side being $x^2 + 2x - 3$ and right-hand side being 0.

An equation has the following properties: 1) any quantity can be added to, subtracted from or multiplied by both sides of the equation;

2) A non-zero quantity can divide both sides; and 3) any function such as power and surds can be applied to both sides equally. After such manipulations, an equation may be converted into a completely different equation, though you can obtain the original equation if you can carefully reverse the above procedure, though special care is needed in some cases.

The idea of these manipulations is to transform the equation to a more simple form whose solution can be found easily.

Example 1.6:   For example, from equation

$$x^2 + 2x - 3 = 0, \tag{1.26}$$

we can add the same term 4 to both sides, and we have

$$x^2 + 2x - 3 + 4 = 0 + 4, \quad \text{or} \quad x^2 + 2x + 1 = 4. \tag{1.27}$$

Since $(x+1)^2 = x^2 + 2x + 1$, we get

$$(x+1)^2 = 4. \tag{1.28}$$

We can take the square root of both sides, we have

$$\sqrt{(x+1)^2} = x + 1 = \pm\sqrt{4} = \pm 2. \tag{1.29}$$

Here we have to keep $\pm$ as the square root of 4 can be either $+2$ or -2 (otherwise, we just find one root). Now we subtract $+1$ from (or add -1 to) both sides, we have

$$x \underbrace{+1 - 1}_{\text{sum}=0} = \pm 2 - 1, \tag{1.30}$$

which leads to

$$x + 0 = \pm 2 - 1, \quad \text{or} \quad x = \pm 2 - 1. \tag{1.31}$$

This is equivalent to moving the term $+1$ on the left to the right and changing its sign so that it becomes -1. Now we have two solutions $x = +2 - 1 = +1$ and $x = -2 - 1 = -3$.

The same mathematical operations can also be applied to a system of equations. In order to solve two simultaneous equations for the unknowns x and y

$$\begin{cases} 5x & +y & = -7 \\ -4x & -2y & = 2 \end{cases}, \tag{1.32}$$

we first multiply both sides of the first equation by 2, we have

$$2 \times (5x + y) = 2 \times (-7), \tag{1.33}$$

and we get

$$10x + 2y = -14. \tag{1.34}$$

Now we add this equation (thinking of the same quantity on both sides) to the second equation of the system. We have

$$\begin{array}{rrl} 10x & +2y & = -14 \\ -4x & -2y & = 2 \\ \hline 6x & +0 & = -12. \end{array} \tag{1.35}$$

This means that

$$6x = -12, \quad \text{or} \quad x = \frac{-12}{6} = -2. \tag{1.36}$$

Now substitute this solution $x = -2$ into one of the equations (say, the first); we have

$$5 \times (-2) + y = -7, \tag{1.37}$$

or

$$-10 + y = -7. \tag{1.38}$$

This simply leads to

$$y = -7 + 10 = 3. \tag{1.39}$$

Now let us look at another example.

Example 1.7: The conversion from Fahrenheit F to Celsius C is given by

$$C = \frac{5}{9}(F - 32).$$

Conversely, from Celsius to Fahrenheit, we have

$$F = \frac{9C}{5} + 32.$$

So the melting point of ice $C = 0°$C is equivalent to $F = \frac{9 \times 0}{5} + 32 = 32°$F, while the boiling point of water $C = 100°$C becomes $9 \times 100/5 + 32 = 212°$F.

We can see that these two temperature scales are very different. However, there is a single temperature at which these two coincide. Let us now try to determine this temperature. We have $C = F$, that is

$$\frac{5}{9}(F - 32) = \frac{9C}{5} + 32 = \frac{9F}{5} + 32,$$

which becomes

$$\left(\frac{5}{9} - \frac{9}{5}\right)F = 32 + \frac{160}{9},$$

or

$$-\frac{56}{45}F = \frac{448}{9}.$$

The solution is simply $F = C = -40$. This means that $-40°$F and $-40°$C represent the same temperature on both scales.

1.3 Index Notation

Before we proceed to study polynomials, let us first introduce the notations of indices.

1.3.1 Notations of Indices

For higher-order products such as $a \times a \times a \times a \times a$ or $aaaaa$, it might be more economical to write it as a simpler form a^5. In general, we use the index form to express the product

$$a^n = \overbrace{a \times a \times ... \times a}^{n}, \qquad (1.40)$$

where a is called the base, and n is called the index or exponent. Conventionally, we write $a \times a = aa = a^2$, $a \times a \times a = aaa = a^3$ and so on and so forth. For example, if $n = 100$, it is obviously advantageous to use the index form. A very good feature of index notation is the multiplication rule

$$a^n \times a^m = \overbrace{a \times a \times ... \times a}^{n} \overbrace{\times a \times a \times ... \times a}^{m} = \overbrace{a \times a \times ... \times a}^{n+m} = a^{n+m}.$$

Thus, the product of factors with the same base can easily be carried out by adding their indices.

If we interpret $a^{-n} = 1/a^n$, we then have the division rule

$$a^n \div a^m = \frac{a^n}{a^m} = a^{n-m}. \qquad (1.41)$$

In the special case when $n = m$, it requires that $a^0 = 1$ because any non-zero number divided by itself should be equal to 1.

If we replace a by $a \times b$ and use $a \times b = b \times a$, we can easily arrive at the factor rule

$$(a \times b)^n = a^n \times b^n = a^n b^n. \qquad (1.42)$$

Similarly, using a^m to replace a in the expression (1.40), it is straightforward to verify the following power-on-power rule

$$(a^m)^n = a^{n \times m} = a^{nm}. \qquad (1.43)$$

Example 1.8: The acceleration, g, due to gravity at the Earth's surface can be calculated using

$$g = \frac{GM_\odot}{R^2},$$

where $G = 6.672 \times 10^{-11}$ m^3 kg^{-1} s^{-2} is the universal gravitational constant, and $R = 6.378 \times 10^3$ km or 6.378×10^6 m is the radius of the

Earth. From the observed standard value of $g \approx 9.80665$ m/s^2, we can estimate the mass of the Earth

$$M_\odot = \frac{gR^2}{G} = \frac{9.80665 \times (6.378 \times 10^6)^2}{6.672 \times 10^{-11}} \approx 5.979 \times 10^{24} \text{ kg}.$$

Here we just mentioned that a is a real number. For any $a > 0$, things are simple. However, what happens when $a < 0$? For example, what does $(-a)^3$ mean? It is easy to extend this if we use the factor rule by using $-a = (-1) \times a$, and we have

$$(-a)^n = (-1 \times a)^n = (-1)^n \times a^n = (-1)^n a^n. \qquad (1.44)$$

This means that $(-a)^n = +a^n$ if n is an even integer (including zero), while $(-a)^n = -a^n$ if n is odd.

All these rules are valid when m, n are integers. In fact, m and n can be fractions or any real numbers. Since $a = a^1 = a^{\frac{1}{n} \times n} = (a^{1/n})^n$, the meaning of $a^{1/n}$ should be

$$a^{\frac{1}{n}} = a^{1/n} = \sqrt[n]{a}. \qquad (1.45)$$

For example, $a^{1/2} = \sqrt{2}$ and $8^{1/3} = \sqrt[3]{8} = \sqrt[3]{2^3} = 2$. In case of a fraction $m = p/q$ where p and $q \neq 0$ are integers, we have

$$a^{p/q} = a^{\frac{p}{q}} = \sqrt[q]{a^p}. \qquad (1.46)$$

This form is often referred to as surds. Similar to integer indices, surds have the following properties:

$$\sqrt[n]{ab} = \sqrt[n]{a} \cdot \sqrt[n]{b}, \quad \sqrt[n]{\sqrt[m]{a}} = \sqrt[nm]{a}, \quad \sqrt[n]{\frac{a}{b}} = \frac{\sqrt[n]{a}}{\sqrt[n]{b}}. \qquad (1.47)$$

Example 1.9: For example, we can simplify the expression

$$\frac{a^5(ab^2)^3}{a^2b^5} = \frac{a^5 a^{1 \times 3} b^{2 \times 3}}{a^2 b^5} = \frac{a^{5+3} b^6}{a^2 b^5} = \frac{a^8}{a^2} \times \frac{b^6}{b^5} = a^{8-2} \times b^{6-5} = a^6 b.$$

Similarly, we have

$$(a^2 \times b^{-2})^{-3} \div \left(\frac{b^5}{a}\right) = a^{2 \times (-3)} \times b^{-2 \times (-3)} \times \frac{a}{b^5}$$

$$= a^{-6} \times b^6 \times \frac{a}{b^5} = a^{-6+1} \times b^{6-5} = a^{-5} b = \frac{b}{a^5}.$$

Let us simplify $\sqrt{-(-4^{1/3})^{6/2} \times (-x^{1/3})^{12}}$. We have

$$\sqrt{-(-4^{1/3})^{6/2} \times (-x^{1/3})^{12}} = [-(-1)^{\frac{6}{2}} \times 4^{\frac{1}{3} \times \frac{6}{2}} \times (-1)^{12} \times x^{\frac{1}{3} \times 12}]^{\frac{1}{2}}$$

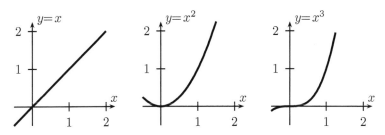

Figure 1.8: Graphs of nth power functions ($n = 1, 2, 3$).

$$= [-1 \times (-1)^3 \times 4 \times 1 \times x^4]^{\frac{1}{2}} = [(-1)^{3+1} \times 4 \times x^4]^{\frac{1}{2}}$$

$$= (2^2 \times x^4)^{\frac{1}{2}} = 2^{2 \times \frac{1}{2}} \times x^{4 \times \frac{1}{2}} = 2x^2.$$

It is worth pointing out that we have a special case when $a = 0$. For $n > 0$, we have $0^n = 0$. However, $n < 0$, it is meaningless in the context we have discussed because, say, $0^{-7} = \frac{1}{0^7}$ involves division by zero. What happens when $a = 0$ and $n = 0$? It should be $0^0 = 1$ for mathematical consistency (though in many standard books, they say 0^0 is also meaningless). Anyway, it is rarely relevant in almost any real-world applications, therefore, we may as well forget it.

Example 1.10: The primary (or P) longitudinal waves and secondary (or S) transverse shear seismic waves have velocities, respectively

$$V_P = \sqrt{\frac{K + \frac{4}{3}G}{\rho}} = \sqrt{\frac{E(1 - \nu)}{\rho(1 + \nu)(1 - 2\nu)}}, \quad V_S = \sqrt{\frac{G}{\rho}} = \sqrt{\frac{E}{2\rho(1 + \nu)}},$$

where K is the bulk modulus and G is the shear modulus of the crust. E is Young's modulus and ν is Poisson's ratio. ρ is the density. If we take the typical values of $E = 80$ GPa $= 8 \times 10^{10}$ Pa, $\nu = 0.25$ and $\rho = 2700$ kg/m^3, we have

$$V_S = \sqrt{\frac{8 \times 10^{10}}{2 \times 2700 \times (1 + 0.25)}} \approx 3400 \text{ m/s},$$

and $V_P \approx 5900$ m/s. In the case of a liquid, the shear modulus is virtually zero ($G \approx 0$), the liquid medium cannot support shear waves, so S-waves cannot penetrate into or pass through fluids. However, both fluids and solids can support P waves. In fact, the absence of S-waves in the Earth's outer core may suggest that it behaves like a liquid.

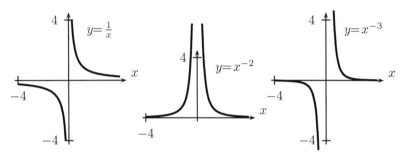

Figure 1.9: Graphs of nth power functions ($n < 0$).

1.3.2 Graphs of Functions

For simple power functions, we can plot them easily, and some examples of positive integer powers of x are shown in Fig. 1.8. For x^n where $n > 0$, all the functions go through two points $(0,0)$ and $(1,1)$. There is no singularity (infinite value caused by the division of zero). When $n < 0$, the graphs are split into different regions and singularity often occurs at some points. For example, functions x^{-1}, x^{-2} and x^{-3} all have a singularity at $x = 0$ because $f(0) = 0^{-n} = 1/0^n \to \infty$ where $n = 1, 2, 3$. In this case, the curves are split into two disconnected branches shown in Fig. 1.9.

A function $f(x)$ is odd if $f(-x) = -f(x)$. For example, $f(x) = x^{-1}$ and $f(x) = x^3$ are odd functions. On the other hand, an even function is a function which satisfies $f(-x) = f(x)$. For example, $f(x) = x^{-2}$ and $f(x) = x^2$ are both even functions. An odd function is rotationally symmetric of $180°$ through the centre or origin at $(0,0)$ because one branch will overlap on the other branch after being rotated by $180°$, while an even function has the reflection symmetry about the y-axis.

However, some functions are neither odd nor even. For example, $f(x) = x^2 - x$ is neither odd nor even because $f(-x) = (-x)^2 - (-x) = x^2 + x \neq f(-x)$ or $f(-x) \neq -f(x)$.

In general, functions can be any form of combinations of all basic functions such as x^n and $\sin x$. In a special case when a function is the sum of many terms and each term consists of x^n only, we are dealing with polynomials, which will be discussed in detail in the next chapter.

1.4 Applications

Before we proceed to introduce more mathematics, let us use a few examples to demonstrate how to apply the mathematical techniques we have learned in earth sciences.

1.4.1   Greenhouse Effect

We now try to carry out an estimation of the Earth's surface temperature assuming that the Earth is a spherical black body. The incoming energy from the Sun on the Earth's surface is

$$E_{\text{in}} = (1 - \alpha)\pi r_E^2 S, \tag{1.48}$$

where α is the albedo or the planetary reflectivity to the incoming solar radiation, and $\alpha \approx 0.3$. S, the total solar irradiance on the Earth's surface, is about $S = 1367$ W/m^2. r_E is the radius of the Earth. Here the effective area of receiving sunlight is equivalent to the area of a disc πr_E^2 as only one side of the Earth is constantly facing the Sun.

A body at an absolute temperature T will have black-body radiation and the total energy E_b emitted by the object per unit area per unit time obeys the Stefan-Boltzmann law

$$E_b = \sigma T^4, \tag{1.49}$$

where $\sigma = 5.67 \times 10^{-8}$ J/K^4 s m^2 is the Stefan-Boltzmann constant. For example, we know a human body has a typical body temperature of $T_h = 36.8°$C or $273 + 36.8 = 309.8$ K. An adult in an environment with a constant room temperature $T_0 = 20°$C or $273 + 20 = 297$ K will typically have a skin temperature $T_s \approx (T_h + T_0)/2 = (36.8 + 20)/2 = 28.4°$C or $273 + 28.4 = 301.4$ K. In addition, an adult can have a total skin surface area of about $A = 1.8$ m^2. Therefore, the total energy per unit time radiated by an average adult is

$$E = A(\sigma T_s^4 - \sigma T_0^2) = A\sigma(T_s^4 - T_0^4)$$

$$= 1.8 \times 5.67 \times 10^{-8} \times (301.4^4 - 297^4) \approx 90 \text{ J/s}, \tag{1.50}$$

which is about 90 watts. This is very close to the power of a 100-watt light bulb.

For the Earth system, the incoming energy must be balanced by the Earth's black-body radiation

$$E_{\text{out}} = A\sigma T_E^4 = 4\pi r_E^2 \sigma T_E^4, \tag{1.51}$$

where T_E is the surface temperature of the Earth, and $A = 4\pi r_E^2$ is the total area of the Earth's surface. Here we have assumed that outer space has a temperature $T_0 \approx 0$ K, though we know from the cosmological background radiation that it has a temperature of about 4 K. However, this has little effect on our estimations.

From $E_{\text{in}} = E_{\text{out}}$, we have

$$(1 - \alpha)\pi r_E^2 S = 4\pi r_E^2 \sigma T_E^4, \tag{1.52}$$

or

$$T_E = \sqrt[4]{\frac{(1-\alpha)S}{4\sigma}}. \tag{1.53}$$

Plugging in the typical values, we have

$$T_E = \sqrt[4]{\frac{(1-0.3) \times 1367}{4 \times 5.67 \times 10^{-8}}} \approx 255 \text{ K}, \tag{1.54}$$

which is about $-18°C$. This is too low compared with the average temperature $9°C$ or 282 K on the Earth's surface. The difference implies that the greenhouse effect of the CO_2 is in the atmosphere. The greenhouse gas warms the surface by about $27°C$.

You may argue that the difference may also come from the heat flux from the lithosphere to the Earth surface, and the heat generation in the crust. That is partly true, but the detailed calculations for the greenhouse effect are far more complicated, and still form an important topic of active research.

1.4.2   Glacier Flow

The dominant mechanism for glacier flow is the power-law creep or viscous flow. The movement depends on factors such as stress, temperature, grain size, impurity and geometry of the flow channel. However, in the simplest case at constant temperature, the flow rate is governed by the Glen flow law in terms a power law relationship between strain rate $\dot{\epsilon}$ and stress τ

$$\dot{\epsilon} = A\tau^n, \tag{1.55}$$

where $n \approx 3$ is the stress exponent and A is constant with a typical mean value of $A \approx 5 \times 10^{-16}$ s^{-1} (kPa)$^{-3}$ for $n = 3$. Glacier movement can be at a very high speed on a geological timescale. For example, it was observed that the Black Rapids Glacier in Alaska can move at the speed of about 40 to 60 m/year. Let us see if this is consistent with the above flow law. We know the stress level τ is about 100 kPa for the glacier, and we have

$$\dot{\epsilon} = A\tau^n \approx 5 \times 10^{-16} \times (100)^3 \approx 5 \times 10^{-10} \text{ s}^{-1}. \tag{1.56}$$

As the transverse range of the glacier is about $w = 3000$ m, the above strain rate is equivalent to the speed

$$v = w\dot{\epsilon} = 3000 \times 5 \times 10^{-10} = 1.5 \times 10^{-6} \text{ m/s}. \tag{1.57}$$

As there are about $a = 365 \times 24 \times 3600 \approx 3.15 \times 10^7$ seconds in a year, the above speed v is approximately

$$v \approx 1.5 \times 10^{-6} \times 3.15 \times 10^7 \approx 47 \text{ m/year}, \tag{1.58}$$

which is just in the right range of actual speed of the glacier movement.

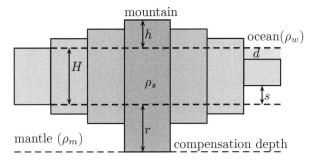

Figure 1.10: Idea of isostasy.

1.4.3 Airy Isostasy

The basic idea of Airy isostasy is essentially the same as the principle
of a floating iceberg. In a mountain area where the height is h, the
root r or the compensation depth of the mountain is related to h. This
means that the crust is thicker in mountain areas.

Since the hydrostatic pressure at a depth h in a fluid of density ρ is
$p = \rho g h$, where g is the acceleration due to gravity, we now consider a
vertical column of a unit area. All pressures below the compensation
depth should be hydrostatic and equal, so we have

$$\rho_s g H + \rho_m g r = \rho_s g(h + H + d), \tag{1.59}$$

where H is the thickness of the crust (see Fig. 1.10). Now we have

$$r = \frac{h \rho_s}{\rho_m - \rho_s}. \tag{1.60}$$

Using the typical values of $\rho_s = 2700$ kg/m^3, and $\rho_m = 3300$ kg/m^3,
we have

$$r = \frac{2700 h}{(3300 - 2700)} = 4.5 h. \tag{1.61}$$

This means that a mountain with $h = 5000$ m would have a root of
$4.5 \times 5000 = 22500$ m or 22.5 km. The surface that separates mantle
from the crust is called Moho, which is almost a reflection or mirror of
the topographical variations.

In the ocean area with the water depth d, the anti-root of the ocean
can also be related to d. The pressures for a vertical column give

$$\rho_s g H + \rho_m g r = \rho_w g d + \rho_s g(H - d - s) + \rho_m g(r + s), \tag{1.62}$$

which gives

$$s = \frac{d(\rho_s - \rho_w)}{\rho_m - \rho_s}. \tag{1.63}$$

Using $\rho_w = 1025$ kg/m^3, we have $s = \frac{(2700 - 1025)d}{(3300 - 2700)} \approx 2.79 d$.

1.4.4 Size of an Impact Crater

The detailed calculations of the size of an impact crater involve many complicated processes; however, we can estimate their approximate size from the kinetic energy of a meteorite or projectile.

For a small meteorite of a mass m with an impact velocity of v_0, we know its kinetic energy E_k is given by $E_k = \frac{1}{2}mv_0^2$. If we assume the meteorite is a sphere with a diameter of D_m and a density of ρ_m, we then have

$$E_k = \frac{1}{2}\rho\frac{\pi D_m^3}{6}v_0^2 = \frac{\pi \rho_m D_m^3 v_0^2}{12}. \tag{1.64}$$

Typically, the velocity is in the range of 11 km/s to 70 km/s and ρ_m can vary significantly. For a stony meteorite, we can use $\rho_m \approx 4000$ km/m^3. Therefore, for a meteorite of a diameter of $D_m = 20$ m travelling at a speed of $v_0 = 15$ km/s=15000 m/s, its kinetic energy is

$$E_k = \frac{3.14 \times 4000 \times (20)^3 \times (15000)^2}{12} \approx 1.88 \times 10^{15} J.$$

If this is converted to the unit of Megatons of equivalent TNT using 1 millon tons of dynamite is equivalent to 4.18×10^{15} J, we have $E_k \approx 1.88 \times 10^{15}/4.18 \times 10^{15} \approx 0.45$ Megatons.

The impact crater of a depth h is created by displacing a volume V of soils with a density of ρ_s. So the potential energy associated with this block of mass is

$$E_p = \rho_s V g h, \tag{1.65}$$

where $g \approx 9.8$ m/s^2 is the acceleration due to gravity. If we further assume that the crater is a hemisphere with a diameter of d_c and $h \approx d_c/2$, we have

$$E_p \approx \rho_s \cdot \frac{\pi d_c^3}{12} \cdot g \frac{d_c}{2} \approx \frac{\pi \rho_s g d_c^4}{24}, \tag{1.66}$$

where we have used the volume of a hemisphere is $V = \frac{2\pi}{3}a^3 = \pi d_c^3/12$ with a being its radius. In reality, only a fraction, γ, of the kinetic energy is used in creating the impact crater and the rest of the energy becomes heat and the energy of the induced shock waves. So the energy balance leads to $E_p = \gamma E_k$. For most impact problems, $\gamma = 0.05$ to 0.2. Thus, the size of the impact crater becomes

$$d_c = \left(\frac{24\gamma E_k}{\pi \rho_s g}\right)^{1/4}. \tag{1.67}$$

If we use $\gamma = 0.05$ and $\rho_s = 2700$ kg/m^3, then the size of the impact crater created by a meteorite of $D_m = 20$ m with $E_k = 1.88 \times 10^{15}$ J can be estimated by $d_c \approx \left(\frac{25\times0.05\times1.88\times10^{15}}{3.14\times2700\times9.8}\right)^{1/4} \approx 400$ m. If we follow the same procedure, a meteorite of $D_m = 100$ m with the same impact velocity will have kinetic energy of about 56 Megatons, which will create a crater of the size $d_c \approx 1.4$ km.

Chapter 2

Preliminary Mathematics II

When we discussed quadratic equations in the previous chapter, we were in fact dealing with a special case of polynomials. Polynomials and their roots or solutions are very important and have applications in many areas in earth sciences. At the end of this chapter, we will use simple examples such as Stokes' flow and raindrop dynamics as our case studies.

2.1 Polynomials

The functions we have discussed so far are a special case of polynomials. In particular, a quadratic function $ax^2 + bx + c$ is a special case of a polynomial. In general, a polynomial $p(x)$ is an explicit expression, often written as a sum or linear combination, which consists of multiples of non-negative integer powers of x. In general, we often write a polynomial in the descending order (in terms of the power x^n)

$$p(x) = a_n x^n + a_{n-1} x^{n-1} + ... + a_1 x + a_0, \qquad (n \geq 0), \qquad (2.1)$$

where $a_i (i = 0, 1, ..., n)$ are known constants. The coefficient a_n is often called the leading coefficient, and the highest power n of x is called the degree of the polynomial. Each partial expression such as $a_n x^n$ and $a_{n-1} x^{n-1}$ is called a term. The term a_0 is called the constant term. Obviously, the simplest polynomial is a constant (degree 0).

When adding (or subtracting) two polynomials, we have to add (or subtract) the corresponding terms (with the same power n of x), and the degree of resulting polynomial may have any degree from 0 to the highest degree of two polynomials. However, when multiplying two polynomials, the degree of the resulting polynomial is the sum of the degrees of the two polynomials. For example,

$$(a_n x^n + a_{n-1} x^{n-1} + ... + a_0)(b_m x^m + b_{m-1} x^{m-1} + ... + b_0)$$

$$= a_n b_m x^{n+m} + ... + a_0 b_0. \tag{2.2}$$

Example 2.1: For example, the quadratic function $3x^2 - 4x + 5$ is a polynomial of degree 2 with a leading coefficient 3. It has three terms.

The expression

$$-2x^5 - 3x^4 + 5x^2 + 1$$

is a polynomial of the fifth degree with a leading coefficient -2. It has only four terms as the terms x^3 and x do not appear in the expression (their coefficients are zero).

If we add $2x^3 + 3x^2 + 1$ with $2x^3 - 3x^2 + 4x - 5$, we have

$$(2x^3+3x^2+1)+(2x^3-3x^2+4x-5)=(2+2)x^3+(3-3)x^2+(0+4)x+(1-5)$$

$$= 4x^3 + 0x^2 + 4x - 4 = 4x^3 + 4x - 4,$$

which is a cubic polynomial. However, if we subtract the second polynomial from the first, we have

$$(2x^3 + 3x^2 + 1) - (2x^3 - 3x^2 + 4x - 5)$$

$$= (2 - 2)x^3 + [3 - (-3)]x^2 + (0 - 4)x + [1 - (-5)]$$

$$= 0x^3 + 6x^2 - 4x + 6 = 6x^2 - 4x + 6,$$

which is a quadratic.

Similarly, the product of $3x^4 - x - 1$ and $-x^5 - x^4 + 2$ is a polynomial of degree 9 because

$$(3x^4 - x - 1)(-x^5 - x^4 + 2)=-3x^9 - 3x^8 + 7x^4 + x^6 + 2x^5 - 2x - 2.$$

- -

2.2 Roots

A natural question now is how to find the roots of a polynomial? We know that there exists an explicit formula for finding the roots of a quadratic. It is also possible for a cubic, and a quartic function (polynomial of degree 4), though quite complicated. In some special cases, it may be possible to find a factor via factorisation or even by educated guess; then some of the solutions can be found. However, in general, factorisation is not possible.

Let us try to solve the following cubic equation

$$f(x) = x^3 - 2x^2 + 3x - 6 = 0,$$

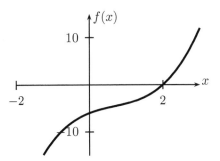

Figure 2.1: Cubic function $f(x) = x^3 - 2x^2 + 3x - 6$.

which can be factorised as

$$x^3 - 2x^2 + 3x - 6 = (x - 2)(x^2 + 3) = 0.$$

So the solution should be either

$$x - 2 = 0,$$

or

$$x^2 + 3 = 0.$$

Therefore, the only real solution is $x = 2$, and the other condition is impossible to satisfy for any real number unless in context of complex numbers, to be discussed later. The graph of $f(x)$ is shown in Fig. 2.1.

Now we introduce a generic method for finding the roots of a general cubic polynomial

$$f(x) = ax^3 + bx^2 + cx + d = 0, \tag{2.3}$$

where a, b, c, d are real numbers, and $a \neq 0$. We can use a simple change of variable

$$z = x + \frac{b}{3a}. \tag{2.4}$$

First, we can write the above equation as

$$f(x) = ax^3 + 3\beta x^3 + 3\gamma x + d = 0, \tag{2.5}$$

where

$$\beta = \frac{b}{3}, \qquad \gamma = \frac{c}{3}. \tag{2.6}$$

Substituting the new variable $z = x + \frac{\beta}{a}$ or $x = z - \frac{\beta}{a}$, we have

$$a(z - \frac{\beta}{a})^3 + 3\beta(z - \frac{\beta}{a})^2 + 3\gamma(z - \frac{\beta}{a}) + d = 0. \tag{2.7}$$

Using the formula $(x+y)^3 = x^3 + 3x^2 y + 3xy^2 + y^3$ and expanding the above equation, after some algebraic manipulations we have

$$z^3 + \frac{3h}{a^2}z + \frac{g}{a^3} = 0, \tag{2.8}$$

where

$$h = a\gamma - \beta^2, \qquad g = a^2 d - 3a\beta\gamma + 2\beta^3. \tag{2.9}$$

This essentially transforms the original equation into a reduced or depressed cubic equation. By defining a constant

$$\delta = f(-b/(3a)) = a(-\frac{b}{3a})^3 + b(-\frac{b}{3a})^2 + c(-\frac{b}{3a}) + d$$

$$= \frac{2b^3}{27a^2} - \frac{bc}{3a} + d = \frac{2\beta^3 - 3a\beta\gamma + a^2 d}{a^2} = \frac{g}{a^2}, \tag{2.10}$$

and the Cardan's discriminant $\Delta = g^2 + 4h^3$, we have

$$\Delta = a^2(a^2 d^2 - 6a\beta\gamma d + 4a\gamma^3 + 4\beta^3 d - 3\beta^2\gamma^2) = a^4(\delta^2 - \epsilon^2), \tag{2.11}$$

or

$$\epsilon^2 = \delta^2 - \frac{\Delta}{a^4}. \tag{2.12}$$

In the case of $\Delta > 0$, the reduced cubic equation has one real root. The real root can be calculated by

$$z = \sqrt[3]{\frac{1}{2a}(-\delta + \sqrt{\delta^2 - \epsilon^2})} + \sqrt[3]{\frac{1}{2a}(-\delta - \sqrt{\delta^2 - \epsilon^2})}, \tag{2.13}$$

or

$$x = z - \frac{b}{3a} = -\frac{b}{3a} + \sqrt[3]{\frac{1}{2a}(-\delta + \frac{1}{a^2}\sqrt{\Delta})} + \sqrt[3]{\frac{1}{2a}(-\delta - \frac{1}{a^2}\sqrt{\Delta})}. \tag{2.14}$$

If $\Delta = 0$, it has three real roots. If $\epsilon \neq 0$, it gives two equal roots, and the three roots are

$$z = \zeta, \ \zeta, \ -2\zeta, \tag{2.15}$$

where ζ is determined by $\zeta^2 = -h/a^2$. Thus, the signs of ζ and δ are important.[1] In modern scientific applications, we rarely use these formulae, as we can easily obtain the solutions using a computer. Here we just use this case to illustrate how complicated the process can become even for cubic polynomials.

---

[1] For details of derivations, please refer to Nickalls, R. W. D., *Mathematical Gazette*, **77**, 354-359 (1993).

Example 2.2:   Now let us revisit the previous example $f(x) = x^3 - 2x^2 + 3x - 6 = 0$. We have $a = 1$, $b = -2$ (or $\beta = -2/3$), $c = 3$ (or $\gamma = 1$), and $d = -6$. Letting $z = x - \frac{b}{3a} = x + 2/3$, we have

$$z^3 + \frac{5}{3}z - \frac{124}{27} = 0.$$

So

$$\delta = \frac{2\beta^3 - 3a\beta\gamma + a^2 d}{a^2} = -\frac{124}{27}, \qquad h = \frac{5}{9},$$

and

$$\Delta = a^2(a^2 d^2 - 6a\beta\gamma d + 4a\gamma^3 + 4\beta^3 d - 2\beta^2\gamma^2) = \frac{196}{9}.$$

Since $\Delta > 0$, there exists a single real root which is given by

$$z = \sqrt[3]{\frac{1}{2}\left(-\frac{-124}{27} + \frac{1}{1^2}\sqrt{\frac{196}{9}}\right)} + \sqrt[3]{\frac{1}{2}\left(-\frac{-124}{27} + \frac{1}{1^2}\sqrt{\frac{196}{9}}\right)}$$

$$= \sqrt[3]{\frac{125}{27}} + \sqrt[3]{\frac{-1}{27}} = \frac{5}{3} + \frac{-1}{3} = \frac{4}{3},$$

where we have used $(-1)^3 = -1$. Therefore, the real root becomes $x = z - \frac{b}{3a} = \frac{4}{3} - \frac{(-2)}{3\times1} = 2$.

- -

For quintic polynomials (degree five) or higher, there is no general formula according to the Abel-Ruffini theorem published in 1824. This means that it is not possible to express the solution explicitly in terms of radicals ($\sqrt[n]{}$ or surds). In this case, approximate methods and numerical methods are the best alternatives.

2.3   Descartes' Theorem

Descartes' rule of signs or theorem can be useful in determining the possible number of real roots. It states that the number of positive roots of a polynomial $p(x)$ is equal to the number of sign changes in the coefficients (excluding zeros) or less than it by a multiple of 2. Similarly, the number of negative roots is the number of sign changes in the coefficients of $p(-x)$ or less than it by a multiple of 2.

Let us demonstrate how it works by using an example. For the polynomial $p(x) = x^3 - 2x^2 - 25x + 50$, the coefficients are $+1, -2, -25$, and $+50$. The number of sign changes of its coefficients are 2 (from $+1$ to -2, and from -25 to $+50$). So the number of positive roots are either 2 or 0 (none). In order to get the number of negative roots, we first have to change the polynomial $p(x)$ to $p(-x)$

$$p(-x) = (-x)^3 - 2(-x)^2 - 25(-x) + 50 = -x^3 - 2x^2 + 25x + 50,$$

whose coefficients are $-1, -2, +25$, and $+50$. So the number of sign changes of the coefficients is 1 (from -2 to $+25$), and the number of negative roots is 1 (it cannot be $1 - 2 = -1$). In fact, the factorisation of $p(x)$ leads to

$$p(x) = (x - 2)(x + 5)(x - 5) = 0,$$

whose roots are $2, -5$, and $+5$. Indeed, there are two positive roots and one negative root.

As another example, let us try the following polynomial

$$f(x) = x^6 - 4x^4 + 3x^2 - 12.$$

The coefficients of x^5, x^3 and x are zeros, so we do not count them. The only valid coefficients are $+1, -4, +3$ and -12. So the number of sign changes is 3, and the number of positive roots is either 3 or $3 - 2 = 1$.

Furthermore, we have

$$f(-x) = (-x)^6 - 4(-x)^4 + 3(-x)^2 - 12 = x^6 - 4x^4 + 3x^2 - 12 = f(x),$$

which is an even function. Therefore, the coefficients are the same, and the number of sign changes is 3. The number of negative roots is either 3 or 1. In fact, the factorisation leads to

$$f(x) = (x + 2)(x - 2)(x^4 + 3),$$

which has a positive root $+2$ and a negative root -2.

Example 2.3:   The permeability k of porous materials such as soil depends on the average grain size $\bar{D}$ and the void ratio ϵ in terms of the well-known Kozeny-Carman relationship

$$k = A\frac{\bar{D}^2 \epsilon^3}{(1 + \epsilon)}, \quad \text{or} \quad \frac{\epsilon^3}{(1 + \epsilon)} = \frac{k}{A\bar{D}^2},$$

where A is a coefficient which depends on the viscosity of the fluid and also slightly on temperature. The unit of k is m^2, though Darcy $\approx 10^{-12}$ m^2 is the common unit in earth sciences and engineering. Suppose we know the values of k, A, and $\bar{D}$ so that $\alpha = \frac{d}{A\bar{D}^2} = \frac{1}{2}$, we want to calculate the void ratio ϵ. Now we have, after multiplying the equation by 2

$$2\epsilon^3 - \epsilon - 1 = 0.$$

Since the signs change once, from Descartes' rule of signs, we know that there is at most one real root and in fact exactly one real root. Since $-\epsilon = \epsilon - 2\epsilon$ and $0 = 2\epsilon^2 - 2\epsilon^2$, we have

$$2\epsilon^3 - \epsilon - 1 = 2\epsilon^3 + 2\epsilon^2 + \epsilon - 2\epsilon^2 - 2\epsilon - 1$$

$$= \epsilon(2\epsilon^3 + 2\epsilon + 1) - (2\epsilon^2 + 2\epsilon + 1) = (\epsilon - 1)(2\epsilon^2 + 2\epsilon + 1) = 0.$$

Therefore, we have either $\epsilon - 1 = 0$, or $2\epsilon^2 + 2\epsilon + 1 = 0$. That is $\epsilon = 1$.

- -

Here we can see that the number of positive and negative roots is just a possible guide, and the actual number has to be determined by other methods. Readers who are interested in such topics can refer to more specialised books.

2.4   Applications

2.4.1   Stokes' Flow

Stokes' law is very important for modelling geological processes such as sedimentation and viscous flow. For a sphere of radius r and density ρ_s falling in a fluid of density ρ_f (see Fig. 2.2), the frictional/viscous resistance or drag is given by Stokes' law

$$F_{\text{up}} = 6\pi\mu v r, \tag{2.16}$$

where μ is the dynamic viscosity of the fluid. v is the velocity of the spherical particle. The driving force F_{down} of falling is the difference between the gravitational force and the buoyant force or buoyancy. That is the difference between the weight of the sphere and the weight of the displaced fluid by the sphere (with the same volume). We have

$$F_{\text{down}} = \frac{4\pi\rho_s g r^3}{3} - \frac{4\pi\rho_f g r^3}{3} = \frac{4\pi(\rho_s - \rho_f)g r^3}{3}, \tag{2.17}$$

where g is the acceleration due to gravity.

The falling particle will reach a uniform velocity v_s, called the terminal velocity or settling velocity, when the drag F_{up} is balanced by F_{down}, or $F_{\text{up}} = F_{\text{down}}$. We have

$$6\pi\mu v_s r = \frac{4\pi(\rho_s - \rho_f)g r^3}{3}, \tag{2.18}$$

which leads to

$$v_s = \frac{2(\rho_s - \rho_f)g r^2}{9\mu} = \frac{(\rho_s - \rho_f)g d^2}{18\mu}, \tag{2.19}$$

where $d = 2r$ is the diameter of the particle.

We know that the typical size of sand particles is about 0.1 mm $= 10^{-4}$ m. Using the typical values of $\rho_s = 2000$ kg/m^3, $\rho_f = 1000$ kg/m^3, $g = 9.8$ m/s^2, and $\mu = 10^{-3}$ Pa s, we have $v_s \approx 0.5 \times 10^{-2}$ m/s $= 0.5$ cm/s. Any flow velocity higher than v_s will result in sand suspension in water and long-distance transport.

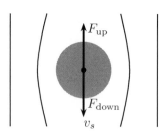

Figure 2.2: Settling velocity of a spherical particle.

Stokes' law is valid for laminar steady flows with very low Reynolds number Re, which is a dimensionless number, and is usually defined as $Re = \rho_f vd/\mu = vd/\nu$, where μ is the viscosity or dynamic viscosity, and $\nu = \mu/\rho_f$ is called the kinematic viscosity. Stokes' law is typically for a flow with $Re < 1$, and such flow is often called the Stokes flow.

From equation (2.19), we can see that if $\rho_s < \rho_f$, then the particle will move up. When you pour some champagne or sparkling water in a clean glass, you will notice a lot of bubbles of different sizes moving up quickly. The size of a bubble will also increase as it moves up; this is due to the pressure decrease and the nucleation process. Large bubbles move faster than smaller bubbles. If we consider a small bubble with negligible change in size, we can estimate the velocity of the bubbles. The dynamic viscosity and density of champagne are about 1.5×10^{-3} Pa s and 1000 kg/m^3, respectively. For simplicity, we can practically assume the density of the bubbles is zero. For a bubble with a radius of $r = 0.1$ mm or diameter $d = 0.2$ mm $= 2 \times 10^{-4}$ m, its uprising velocity can be estimated by

$$v_s = \frac{(1000 - 0) \times 9.8(2 \times 10^{-4})^2}{18 \times 1.5 \times 10^{-3}} \approx 0.015 \text{ m/s} = 1.5 \text{ cm/s}. \quad (2.20)$$

Similarly, gas bubbles in hot magma or a volcano also rise. The viscosity of magma varies widely from 10^2 to 10^{17} Pa s, depending on the temperature and composition. For granitic liquid, its typical viscosity is $\eta = 10^5$ Pa s with a density of $\rho_s = 2700$ kg/m^3. A gas bubble with a diameter $d = 1$ cm $= 0.01$ m, its ascent velocity is

$$v_s = \frac{(2700 - 0) \times 9.8 \times (0.01)^2}{18 \times 10^5} \approx 1.5 \times 10^{-6} \text{ m/s} \approx 47 \text{ m/year}, \quad (2.21)$$

which means that it will take more than 500 years for a bubble of this size to rise from 25km deep to the surface.

2.4.2 Velocity of a Raindrop

Raindrops vary in size from about 0.1 mm to 5.5 mm. Now the fluid is the air with density $\rho_a = 1.2$ kg/m^3, and viscosity $\mu_a = 1.8 \times 10^{-5}$ Pa s. For a very small cloud drop or raindrop $d = 0.15$ mm $= 1.5 \times 10^{-4}$ m with a density of $\rho_w = 1000$ kg/m^3, we can estimate its terminal velocity as

$$v = \frac{(\rho_w - \rho_a)gd^2}{18\mu_a} = \frac{(1000 - 1.2) \times 9.8 \times (1.5 \times 10^{-4})^2}{18 \times 1.8 \times 10^{-5}} \approx 0.68 \text{ m/s},$$

which is about the same value as observed by experiment. In this case, the Reynolds number is approximately $Re = \frac{\rho_a vd}{\mu_a} \approx 6.8$, which is bigger than 1. There will be some difference between estimated values and the real velocity.

However, for larger raindrops, their falling velocities are high, and Stokes' law is no longer valid. We have to use a different approach to calculate the terminal velocity. Stokes' law should be modified to

$$F_{\text{drag}} = \frac{1}{2}\rho_a C_d v^2 A, \tag{2.22}$$

where C_d is a dimensionless drag coefficient which depends on the shape and Reynolds number of the falling particle. For raindrops, it is typically in the range of 0.4 to 0.8 for small spherical raindrops to large raindrops. Here A is the perpendicular area, and $A = \pi r^2 = \pi d^2/4$ for a sphere. So the drag force is

$$F_{\text{drag}} = \frac{1}{2}C_d\rho_a v^2 \cdot \frac{\pi d^2}{4} = \frac{1}{8}\pi C_d\rho_a v^2 d^2. \tag{2.23}$$

The downward force is given by

$$F_{\text{down}} = \frac{4\pi\rho_w gr^3}{3} - \frac{4\pi\rho_a gr^3}{3} = \frac{\pi(\rho_w - \rho_a)gd^3}{6}, \tag{2.24}$$

where we have used $d = 2r$. Now the terminal velocity is determined by the balance $F_{\text{drag}} = F_{\text{down}}$ or

$$\frac{1}{8}\pi C_d\rho_a v^2 d^2 = \frac{\pi(\rho_w - \rho_a)gd^3}{6}, \tag{2.25}$$

which gives

$$v = \sqrt{\frac{4(\rho_w - \rho_a)gd}{3C_d\rho_a}}. \tag{2.26}$$

The viscosity does not appear in the expression explicitly; however, the terminal velocity does depend on viscosity through the implicit dependence of C_d on Re which includes viscosity. For large raindrops, the Reynolds numbers are typically in the range of $Re = 200$ to 1500.

The largest raindrops typically have a diameter of $d = 5.5$ mm$=5.5 \times 10^{-3}$ m, and $C_d = 0.5$. The terminal velocity is approximately

$$v_\infty = \sqrt{\frac{4 \times (1000 - 1.2) \times 9.8 \times d}{3 \times 0.5 \times 1.2}} \approx \sqrt{2.2 \times 10^4 d}. \qquad (2.27)$$

Substituting $d = 5.5 \times 10^{-3}$m, we have $v_s \approx \sqrt{2.2 \times 10^4 \times 5.5 \times 10^{-3}}$ ≈ 11 m/s, which is not far from the measured velocity $v_\infty \approx 9$ to 12 m/s by Beard.[2] This speed range is equivalent to 33 km per hour (or 20 mph) to 43 km/h (or 27 mph). Such an impact velocity will provide enough energy for severe soil erosion.

As we can see, larger raindrops fall faster than smaller ones, so they will catch up and collide with many smaller raindrops on the way, and thus larger raindrops will ultimately break by such frequent bombardment before getting too large. This limits the size of the largest possible raindrops to about 5.5 mm to 7 mm.

However, for hailstones, their sizes are much larger and thus their terminal velocities are higher according to (2.27). For example, a hailstone of $d=1$ cm$= 0.01$ m would typically have a terminal velocity $v_s = \sqrt{2.2 \times 10^4 \times 0.01} \approx 15$ m/s, or 54 km/h. For $d = 5$ cm, $v_s \approx 33$ m/s or 119 km/h. It could cause a lot of damage.

Snowflakes drift down slowly and their terminal velocities can also be estimated by equation (2.26). The density of snowflakes is typically 5% to 12% of the density of water, or $\rho_s = 50$ to 120 kg/m^3 depending on the weather conditions. The size of a snowflake can usually vary from 1 mm to 3 cm or even larger, though most are a few millimetres in diameter. Using the values of $\rho_s = 80$ kg/m^3, $\rho_a = 1.2$ kg/m^3 for air, $D = 3$ mm $= 3 \times 10^{-3}$ m, and the drag coefficient $C_D = 1.15$ for a disk, we have

$$v_s = \sqrt{\frac{4 \times (80 - 1.2) \times 9.8 \times 4 \times 10^{-3}}{3 \times 1.2 \times 1.15}} \approx 1.5 \text{ m/s}, \qquad (2.28)$$

which is about 5.4 km/h.

---

[2]Bead, K. V., Terminal velocity and shape of cloud and precipitation drops aloft, *J. Atmos. Sci.*, **273**, 1359-1364 (1076).

Chapter 3

Binomial Theorem and Sequences

After reviewing the fundamentals of the polynomials and index notations, we are now ready to introduce more important topics in mathematics. In the notations of indices, we have tried to simplify certain expressions such as $(a^4b^2)^5/a^{12}$, but we never actually dealt with the factors such as $(a+b)^5$. This requires the introduction of the binomial theorem, which is an important mathematical technique. Binomial theorem is a basis for mathematical analysis of probability as commonly used in geostatistics and data modelling. Sum of series is crucial for numerical analysis and approximation techniques.

In earth sciences, we often have to deal with powers and logarithms. Logarithms become very convenient for dealing with numbers with a wide range of variations, as we will demonstrate in describing the scale and distributions of earthquakes and near-Earth objects.

3.1 The Binomial Theorem

From the discussions in the previous chapters, we know $(a+b)^0 = 1$ and $(a+b)^1 = (a+b) = a+b$. We also know that

$$(a+b)^2 = (a+b)(a+b) = a(a+b) + b(a+b) = a^2 + 2ab + b^2. \quad (3.1)$$

If we follow a similar procedure to expand, we can deal with higher-order expansions. For example,

$$(a+b)^3 = (a+b)(a+b)^2 = (a+b)(a^2 + 2ab + b^2),$$

$$= a(a^2 + 2ab + b^2) + b(a^2 + 2ab + b^2) = a^3 + 3a^2b + 3ab^2 + b^3. \quad (3.2)$$

Similarly, we have

$$(a+b)^4 = a^4 + 4a^3b + 6a^2b^2 + 4ab^3 + b^4, \tag{3.3}$$

$$(a+b)^5 = a^5 + 5a^4b + 10a^3b^2 + 10a^2b^3 + 5ab^4 + b^5. \tag{3.4}$$

These expressions become longer and longer, and now the question is whether the general expression exists for $(a+b)^n$ for any integer $n > 0$? The answer is yes, and this is the binomial theorem.

3.1.1   Factorials

Before we introduce the binomial theorem, we have to introduce some other relevant concepts such as factorials.

For any positive integer n, the factorial n (or n factorial), denoted by $n!$, is the product of all the n natural numbers from 1 to n. That is

$$n! = n \times (n-1) \times (n-2) \times \ldots \times 1. \tag{3.5}$$

For example, we have $1! = 1$, $2! = 2 \times 1 = 2$, $3! = 3 \times 2 \times 1 = 6$, and $5! = 5 \times 4 \times 3 \times 2 \times 1 = 120$. Similarly, $10! = 10 \times 9 \times \ldots \times 2 \times 1 = 3628800$. From the above definition, we can easily obtain a recursive formula

$$(n+1)! = n! \times (n+1), \qquad (n = 1, 2, 3, \ldots). \tag{3.6}$$

In fact, the above equation is still valid for $n = 0$. Now you may wonder what is the value of $0!$? For consistency and simplicity in writing mathematical expressions, it requires that we define $0! = 1$. In fact, it is also possible to define the factorial of negative integers, but it involves more difficult concepts such as Γ function and Γ integrals. Anyway, factorials of negative numbers such as $(-5)!$ are rarely used in earth sciences, so we do not need to study them further.

The combinatorial coefficient or binomial coefficient is often written in the form

$$\binom{n}{r} \equiv {}^nC_r \equiv \frac{n!}{r!(n-r)!}, \tag{3.7}$$

where $n, r \geq 0$ are integers. Here the symbol '$\equiv$' means 'it is defined as' or 'exactly the same as'. In some literature, the notation ${}^nC_r$ is also widely used. To write the expression explicitly, we have

$$\binom{n}{r} = \frac{n(n-1)(n-2)\ldots(n-r+1)}{1 \times 2 \times \ldots \times (r-1) \times r}. \tag{3.8}$$

For example, we can calculate $\binom{10}{5}$ by using either $\binom{10}{5} = \frac{10 \times 9 \times 8 \times 7 \times 6}{1 \times 2 \times 3 \times 4 \times 5}$ $= \frac{30240}{120} = 252$, or $\binom{10}{5} = \frac{10!}{5!(10-5)!} = \frac{3628800}{120 \times 120} = 252$.

Special cases are when $r = 0$ and $r = n$, we have $\binom{n}{0} = 1$, and $\binom{n}{n} = 1$. From the definition of a factorial and (3.6), we have

$$\binom{n}{m+1} = \frac{n!}{(m+1)!(n-m-1)!}$$

$$= \frac{n! \times (n-m)}{m! \times (m+1) \times \underbrace{(n-m-1)! \times (n-m)}_{=(n-m)!}}$$

$$= \frac{n!}{m!(n-m)!} \times \frac{(n-m)}{m+1} = \frac{(n-m)}{(m+1)} \binom{n}{m}, \qquad (3.9)$$

which is valid for $m = 0, 1, 2, \ldots$ and $n \geq m$.

3.1.2   Binomial Theorem and Pascal's Triangle

Now we can state the binomial theorem, for any natural number n, as

$$(a+b)^n = \binom{n}{0}a^n + \binom{n}{1}a^{n-1}b + \binom{n}{2}a^{n-2}b^2 + \ldots + \binom{n}{n-1}ab^{n-1} + \binom{n}{n}b^n.$$

Example 3.1:   In the case when $n = 5$, the binomial theorem gives

$$(a+b)^5 = \binom{5}{0}a^5 + \binom{5}{1}a^4b + \binom{5}{2}a^3b^2 + \binom{5}{3}a^2b^4 + \binom{5}{4}ab^4 + \binom{5}{5}b^5.$$

It is straightforward to verify that

$$\binom{5}{0} = 1, \quad \binom{5}{1} = 5, \quad \binom{5}{2} = 10, \quad \binom{5}{3} = 10, \quad \binom{5}{4} = 5, \quad \binom{5}{5} = 1,$$

and we have $(a+b)^5 = a^5 + 5a^4b + 10a^3b^2 + 10a^2b^3 + 5ab^4 + b^5$.

We can see that there is a certain symmetry in the coefficients. In fact, the coefficients form the famous pattern – Pascal's triangle.

$$
\begin{array}{ccccccccccc}
 & & & & & 1 & & & & & \\
 & & & & 1 & & 1 & & & & \\
 & & & 1 & & 2 & & 1 & & & \\
 & & 1 & & 3 & & 3 & & 1 & & \\
 & 1 & & 4 & & 6 & & 4 & & 1 & \\
1 & & 5 & & 10 & & 10 & & 5 & & 1
\end{array}
$$

$$\cdots \cdots$$

The coefficients on the extreme left (and right) are always 1, and the coefficient of the next row is the sum of the two coefficients on its 'shoulders' in the previous row. For example, 10 in the 5th row is the sum of 4 and 6 in the 4th row, so is $5 = 4 + 1$.

In general, we have the recursive relationship

$$\binom{n}{r} = \binom{n-1}{r-1} + \binom{n-1}{r}, \qquad (n = 1, 2, 3, ...). \qquad (3.10)$$

Example 3.2: We now use (3.9) to prove the above relationship. We know that

$$\binom{n-1}{r} = \binom{n-1}{(r-1)+1} = \frac{(n-1)-(r-1)}{(r-1)+1}\binom{n-1}{r-1} = \frac{n-r}{r}\binom{n-1}{r-1},$$

we then have

$$\binom{n-1}{r-1} + \binom{n-1}{r} = \binom{n-1}{r-1} + \frac{(n-r)}{r}\binom{n-1}{r-1}$$

$$= \frac{r+(n-r)}{r}\binom{n-1}{r-1} = \frac{n}{r}\binom{n-1}{r-1}$$

$$= \frac{n}{r} \times \frac{(n-1) \times (n-2) \times ... \times ((n-1)-(r-1)+1)}{1 \times 2 \times ... \times (r-1)}$$

$$= \frac{n \times (n-1) \times ... \times (n-r+1)}{1 \times 2 \times ... \times (r-1) \times r} = \binom{n}{r}.$$

An interesting issue here is that you may wonder what happens if we extend the integer n to fractions and even negative numbers, say $(a+b)^{-1/2}$? This will lead to the generalised binomial theorems which involve infinite series.

The above formula provides a way to construct the coefficients from $n = 1$. In essence, this formula describes a sequence.

3.2 Sequences

The simplest sequence is probably the natural numbers

$$1, \ 2, \ 3, \ 4, \ 5, \ ... \qquad (3.11)$$

where we know that the next numbers should be 6, 7 and so on and so forth. Each isolated number is called a term. So 1 is the first term, and 5 is the 5th term. Mathematically speaking, a sequence is a row of numbers that obeys certain rules, and such rules can either be a formula, a recursive relationship (or equation) between consecutive terms, or any other deterministic or inductive relationship. In general, we have a sequence

$$a_1, \quad a_2, \quad a_3, \quad ..., \quad a_n, ... \qquad (3.12)$$

where a_1 is the first term and the nth term is a_n. It is worth pointing out that it is sometimes more convenient to write the sequence as $a_0, a_1, a_2, a_3, ..., a_n, ...$, depending on the context. In this case, we have to be careful to name the term. The first term should be a_0 and the nth term is a_{n-1}.

Example 3.3: For example, the sequence

$$2, \ 5, \ 10, \ 17, \ 26, \ ..., \qquad (3.13)$$

can be described by the formula for the nth term

$$a_n = n^2 + 1, \qquad (n = 1, 2, 3, ...).$$

The sequence $-1, \ 1, \ 3, \ 5, \ 7, \ ...$ corresponds to the formula $a_n = 2n - 3$ for $(n = 1, 2, 3, ...)$. It can also be described by the recursive formula

$$a_{n+1} = a_n + 2, \qquad a_1 = -1, \qquad (n = 1, 2, 3, ...).$$

There is a constant difference between consecutive terms a_n and a_{n+1}.
    On the other hand, the sequence

$$2, \ 1, \ \frac{1}{2}, \ \frac{1}{4}, \ \frac{1}{8}, \ ...$$

has a constant ratio $r = 1/2$. The general formula is

$$a_n = 2r^{n-1}, \qquad (n = 1, 2, 3, ...),$$

which is also equivalent to the recursive relationship

$$a_{n+1} = a_n r, \qquad a_1 = 2, \ (n = 1, 2, 3, ...).$$

- -

There are two special classes of widely-used sequences. They are arithmetic sequences and geometric sequences. In the previous example, the formula $a_n = 2n - 3$ $(n = 1, 2, 3, ...)$ is an arithmetic sequence because there is a common difference $d = 2$ between a_{n+1} and a_n. On the other hand, the formula $a_n = 2r^{n-1}$ describes a geometric sequence with a common ratio r.
    In general, an arithmetic sequence, also called arithmetic progression, can be written as

$$a_n = a + (n - 1)d, \qquad (3.14)$$

where a is a known constant (often the first term), and d is the common difference (also called the step). We often assume $d \neq 0$, otherwise we have a trivial sequence with every term being the same number. For example, the formula $a_n = 2n - 3$ for the sequence $-1, 1, 3, 5, ...$ can be written as

$$a_n = 2n - 3 = -1 + 2(n - 1), \tag{3.15}$$

which gives $a = -1$ and $d = 2$.

On the other hand, a geometric sequence, or geometric progression, is defined by $a_1 = a$ and $a_{n+1} = ra_n(n = 1, 2, 3, ...)$, where the common ratio $r \neq 0$ or 1. This definition gives

$$a_1 = a, \ a_2 = ar, \ a_3 = ar^2, \ ..., \ a_n = ar^{n-1}, \ ..., \tag{3.16}$$

whose nth term is simply

$$a_n = ar^{n-1}, \qquad (n = 1, 2, 3, ...). \tag{3.17}$$

Obviously, r can be negative. For example, for $a = 1$ and $r = -\frac{2}{3}$, we have

$$1, \ -\frac{2}{3}, \ \frac{4}{9}, \ -\frac{9}{27}, \ \frac{16}{81}, \ ..., \tag{3.18}$$

whose nth term is

$$a_n = (-\frac{2}{3})^{n-1}. \tag{3.19}$$

3.2.1 Fibonacci Sequence

Now the question is which form of formula we should use? In general, we should use the formula for the nth term because we can easily calculate the actual number for each term. If we try to study the relationship between terms (often among consecutive terms, though not always), we should use the recursive relationship. It is worth pointing out that the relationship between these two forms could be complicated in some cases. Let us look at a classic example.

Example 3.4: The famous classic sequence is the Fibonacci sequence

$$0, \ 1, \ 1, \ 2, \ 3, \ 5, \ 8, \ 13, \ 21, ...$$

whose recursive relationship (or equation) is

$$a_{n+1} = a_n + a_{n-1}, \qquad (n = 1, 2, 3, ...),$$

with $a_0 = 0$ and $a_1 = 1$. The formula for the nth term is not simple at all, and its construction often involves the solution to a recurrence equation.

- -

In order to get the general expression of nth term for the Fibonacci sequence, we have to solve its corresponding recurrence equation

$$a_{n+1} = a_n + a_{n-1}. \tag{3.20}$$

This is a second-order equation because it involves the relationship among three consecutive terms. First, we can assume that the general term consists of a generic factor $a_n = B\lambda^n$ where λ is the unknown value to be determined, and B is an arbitrary constant.[1]

Substituting $a_n = B\lambda^n$, equation (3.20) becomes

$$B\lambda^{n+1} = B\lambda^n + B\lambda^{n-1}. \tag{3.21}$$

By dividing $B\lambda^{n-1}$ (assuming it is not zero, otherwise the solution is trivial, that is zero), we have a characteristic equation

$$\lambda^2 - \lambda - 1 = 0. \tag{3.22}$$

This is a quadratic equation for λ, and its solution can be obtained easily. We have

$$\lambda_1 = \frac{1 + \sqrt{5}}{2}, \qquad \lambda_2 = \frac{1 - \sqrt{5}}{2}. \tag{3.23}$$

The trick here is again to assume that the generic form of the solution is a linear combination of the two possible basic solutions

$$a_n = \alpha\lambda_1^n + \beta\lambda_2^n = \alpha(\frac{1+\sqrt{5}}{2})^2 + \beta(\frac{1-\sqrt{5}}{2})^n, \tag{3.24}$$

where α and β are undetermined coefficients. Now we have to determine α and β using the initial two terms $a_0 = 0$ and $a_1 = 1$. For $n = 0$, we have

$$a_0 = 0 = \alpha + \beta, \tag{3.25}$$

which means $\beta = -\alpha$. For $n = 1$, we have

$$a_1 = 1 = \alpha(\frac{1+\sqrt{5}}{2}) + \beta(\frac{1-\sqrt{5}}{2}) = \alpha(\frac{1+\sqrt{5}}{2}) - \alpha(\frac{1-\sqrt{5}}{2}). \tag{3.26}$$

Its solution is simply $\alpha = 1/\sqrt{5}$. Subsequently, we have $\beta = -\alpha = -1/\sqrt{5}$. Therefore, the general formula for the nth term in the Fibonacci sequence is

$$a_n = \frac{1}{\sqrt{5}}[(\frac{1+\sqrt{5}}{2})^n - (\frac{1-\sqrt{5}}{2})^n]. \tag{3.27}$$

This formula is almost impossible to guess from the relationship of the sequence itself. It leaves us with the task of verifying that this formula indeed provides the right terms for the Fibonacci sequence.

---

[1]Readers can refer to more advanced textbooks such as Yang, X. S., *Mathematical Modelling for Earth Sciences*, Dunedin Academic, (2008) and its bibliography.

3.2.2   Sum of a Series

Sometimes, we have to calculate the sum of all the terms of a sequence. For example, when Gauss was a child, he was able to add all the natural numbers from 1 to 100 in an amazingly short time.

$$1 + 2 + 3... + 99 + 100 = 5050. \tag{3.28}$$

This is because the sum of all the numbers from 1 to n is

$$1 + 2 + ... + n = \frac{n(n+1)}{2}. \tag{3.29}$$

In this case, we are in fact dealing with the sum of all the terms of a sequence. Traditionally, we often refer to a sequence as a series when we are dealing with the sum of all the terms (usually, a finite number of terms).

Conventionally, we use the sigma notation, $\sum$ from the Greek capital letter S, to express the sum of all the terms in a series

$$\sum_{i=1}^{n} a_i = a_1 + a_2 + ... + a_n, \tag{3.30}$$

where a_i denotes the general term, and the sum starts from $i = 1$ until the last term $i = n$.

For the sum of an arithmetic series with the ith term of $a_i = a + (i-1)d$, we have

$$S = \sum_{i=1}^{n} a_i = a_1 + a_2 + ... + a_n = \underbrace{a}_{=a_1} + (a+d) + (a+2d) + ... + \underbrace{(a + (n-1)d)}_{=a_n}$$

$$= a_1 + (a_1 + d) + (a_1 + 2d) + ... + (a_n - d) + a_n. \tag{3.31}$$

We can also write this sum from the back (last term a_n) to the front (first term $a_1 = a$), and we have

$$S = a_n + (a_n - d) + (a_n - 2d) + ... + (a_n - (n-2)d) + a$$

$$= a_n + (a_n - d) + (a_n - 2d) + ... + (a_1 + 2d) + (a_1 + d) + a_1. \tag{3.32}$$

Adding these above two expressions and grouping the similar terms, we have

$$2S = (a_1 + a_n) + (a_1 + d + a_n - d) + ... + (a_n - d + a_1 + d) + (a_n + a_1)$$

$$= (a_1 + a_n) + (a_1 + a_n) + ... + (a_1 + a_n) = n(a_1 + a_n). \tag{3.33}$$

Therefore, the sum S is

$$S = \frac{n}{2}(a_1 + a_n). \tag{3.34}$$

As $a_1 = a$ and $a_n = a + (n-1)d$, we also have

$$S = \frac{n}{2}[2a + (n-1)d] = n[a + \frac{(n-1)}{2}d].\qquad(3.35)$$

Example 3.5: In the case of the sum of all natural numbers $1 + 2 + 3 + ... + n$, we have $a = 1$ and $d = 1$, so the sum is

$$S = 1 + 2 + 3 + ... + n = \frac{n}{2}[2 \times 1 + (n-1) \times 1] = \frac{n(n+1)}{2}.$$

For the sum of $S = 1 + 3 + 5 + ... + (2n-1)$, we have $a = 1$, $d = 2$ and $a_n = 2n - 1$. Now we have

$$S = 1 + 3 + 5 + ... + (2n-1) = \frac{n}{2}[1 + (2n-1)] = n^2.$$

- -

For the sum of a geometric series, we have

$$S = \sum_{i=1}^{n} a_i = \overbrace{a + ar + ar^2 + ... + ar^{n-2} + ar^{n-1}}.\qquad(3.36)$$

Multiplying the above equation by r on both sides, we have

$$rS = \overbrace{ar + ar^2 + ar^3 + ... + ar^{n-2} + ar^{n-1}} + ar^n.\qquad(3.37)$$

Now if we subtract the first from the second equation, we have

$$rS - S = ar^n - a,\qquad(3.38)$$

or

$$(r-1)S = a(r^n - 1),\qquad(3.39)$$

which leads to

$$S = \frac{a(r^n - 1)}{(r-1)} = \frac{a(1 - r^n)}{(1-r)}.\qquad(3.40)$$

It is worth pointing out that it may cause problems when $r = 1$ becomes the division by zero. In fact, when $r = 1$, all the terms in the series are the same (as the first term a), so the sum is now $a + a + ... + a = na$.

Another special case is when $r = -1$; then we have an oscillatory series $a, -a, +a, -a, ...$, and the sum will be $S = 0$ when n is even, or $S = a$ when n is odd. Both $r = 1$ and $r = -1$ are thus trivial and we will not discuss them any further.

Example 3.6: For the sum of the geometric series

$$S = 1 + (\frac{-2}{3}) + (\frac{-2}{3})^2 + ... + (\frac{-2}{3})^{n-1},$$

we have $a = 1$ and $r = -2/3$, so we have

$$S = \frac{1 \times [1 - (\frac{-2}{3})^n]}{1 - (-\frac{2}{3})} = \frac{1 - (\frac{-2}{3})^n}{\frac{5}{3}}.$$

For example, when $n = 10$, we have

$$S = 1 - \frac{2}{3} + \frac{4}{9} - \frac{9}{27} + \dots - \frac{512}{19683} = \frac{1 - (-2/3)^{10}}{5/3} = \frac{11605}{19683}.$$

- -

Because the addition of any two numbers is interchangeable, that is $a + b = b + a$, it is easy to verify that the sum of two series follows the addition rule

$$\sum_{i=1}^{n}(a_i + b_i) = (a_1 + b_1) + (a_2 + b_2) + \dots + (a_n + b_n)$$

$$= (a_1 + \dots + a_n) + (b_1 + \dots + b_n) = \sum_{i=1}^{n} a_i + \sum_{i=1}^{n} b_i. \qquad (3.41)$$

Similarly, for any non-zero constant β, we have

$$\sum_{i=1}^{n} \beta a_i = \beta a_1 + \beta a_2 + \dots + \beta a_n = \beta(a_1 + \dots + a_n) = \beta \sum_{i=1}^{n} a_i. \quad (3.42)$$

These properties become useful in calculating the sum of complicated series. Let us look at an example.

Example 3.7: We now try to calculate the sum of the series

$$S = 5 + 10 + 18 + 32 + 58 + 108 + \dots + 1556.$$

The general formula to generate this sequence is

$$a_n = 2n + 3 \times 2^{n-1}.$$

The sum can be written as

$$S = \sum_{i=1}^{n} a_i = \sum_{i=1}^{n}(2i + 3 \times 2^{i-1}) = 2\sum_{i=1}^{n} i + 3\sum_{i=1}^{n} 2^{i-1}.$$

The first part is twice the sum of an arithmetic series, while the second part is the sum of a geometric series multiplied by a factor of 3. Using the sum formula for arithmetic and geometrical series, we have the sum

$$S = 2 \times \frac{n(n+1)}{2} + 3 \times \frac{1 \times (2^n - 1)}{(2-1)} = n(n+1) + 3(2^n - 1).$$

In the case of $n = 10$, we have

$$S = 5 + 10 + 18 + 32 + \dots + 1556 = 10 \times (10 + 1) + 3 \times (2^{10} - 1) = 3179.$$

- -

3.2.3   Infinite Series

For an arithmetic series, the sum $S = n[a + \frac{n-1}{2}d]$ will increase as n increases for $d > 0$, and this increase will be unbounded (to any possible large number).

On the other hand, the sum of a geometrical series

$$S = \frac{a(1 - r^n)}{1 - r},\tag{3.43}$$

will have interesting properties as n increases to very large values. For $r > 1$, r^n will increase indefinitely as n tends to infinity, and the magnitude or the absolute value $|S|$ of the sum S will also increase indefinitely. In this case, we say the sum diverges to infinity. Similarly, for $r < -1$, r^n also increase indefinitely (to large negative when n is odd, or to large positive when n is even). As mentioned earlier, both $r = 1$ and $r = -1$ are trivial cases. So we are now only interested in the case $-1 < r < 1$, or $|r| < 1$ where $|r|$ is the absolute value of r. For example, $|-0.5| = 0.5$ and $|0.5| = 0.5$.

As n increases and tends to infinity, we use the notation $n \to \infty$. In the case of $-1 < r < 1$, the interesting feature is that r^n becomes smaller and smaller, approaching zero. We denote this by $r^n \to 0$ as $n \to \infty$. As the number of terms in the series increases to infinity, we call the series an infinite series. Therefore, the sum of an infinite geometrical series with $|r| < 1$ will tend to a finite value (the limit), and we will use the following notation

$$S = a + ar + ar^2 + ... + ar^{i-1} + ...$$

$$= \lim_{n \to \infty} \sum_{i=1}^{n} a_i = \lim_{n \to \infty} \frac{a(1 - r^n)}{1 - r} = \frac{a}{1 - r}.\tag{3.44}$$

Let us look at some applications.

Example 3.8:   The classic example is the sum of

$$S = 1 + \frac{1}{2} + \frac{1}{4} + \frac{1}{8} + \frac{1}{16} + ...,$$

where $a = 1$ and $r = 1/2$. So we have

$$S = \frac{a}{1 - r} = \frac{1}{1 - 1/2} = 2.$$

For the infinite series

$$S = 3 - 3 \cdot \frac{1}{5} + 3 \cdot \frac{1}{25} - 3 \cdot \frac{1}{125} + ...,$$

we have to change the series into

$$S = 3 \times [1 - \frac{1}{5^1} + \frac{1}{5^2} - \frac{1}{5^3} + ...] = 3 \sum_{i=1}^{n} (\frac{-1}{5})^{i-1} = 3 \times \frac{1}{1 - (\frac{-1}{5})} = \frac{5}{2}.$$

3.3   Exponentials and Logarithms

When we discussed the geometric sequence, we frequently used the nth term ar^{n-1} where $a \neq 0$ and the ratio r are given constants. By varying the positive integer n, we can calculate the value of any term in the sequence. If we extend the values of n from integers to real numbers x, the function r^x is in fact an exponential function.

So an exponential function is in the generic form b^x

$$f(x) = b^x, \tag{3.45}$$

where $b > 0$ is a known or given real constant, called the base, while the independent variable x is called the exponent. In the context of real numbers, we have to limit our discussion here to $b > 0$. When $b < 0$, for some values of x, b^x will be meaningless. For example, if $b = -3$ and $x = 1/2$, then $(-3)^{1/2} = \sqrt{-3}$ would be meaningless. Similarly, say, when $b = 0$ and $x = -5$, then $b^x = 0^{-5} = 1/0^5$ would also be meaningless.

In addition, a trivial case is when $b = 1$, then $f(x) = 1^x = 1$ everywhere. It becomes a single constant, not a function in the standard sense. Three functions $f(x) = 2^x, (\frac{1}{2})^x$ and $(\frac{1}{4})^x$ are shown in Fig. 3.1 where all the functions will go though the points $(0, 1)$ where $x = 0$ and $f(0) = 1$. If $b > 0$, then $b^x > 0$ for all x. This is because if $x \geq 0$, $b^x \geq 1 > 0$, while $x < 0$, $b^x = b^{-|x|} = \frac{1}{b^{|x|}} > 0$. We can also see that the reflection in the y−axis of 2^x is $(1/2)^x$. In fact, both b^x and $(1/b)^x$ are the reflection of each other in the y−axis.

An interesting extension for $f(x) = a^x$ is that

$$f(-x) = a^{-x} = \frac{1}{a^x} = (\frac{1}{a})^x. \tag{3.46}$$

Therefore, the division of two exponential functions a^x and b^x is

$$\frac{a^x}{b^x} = (\frac{a}{b})^x. \tag{3.47}$$

Now suppose we know the value of $b > 0$ and the value of the function $y = b^x$, a challenging question is if we can determine x given b and y. This is essentially to calculate the inverse of the exponential function so as to find x that satisfies

$$b^x = y. \tag{3.48}$$

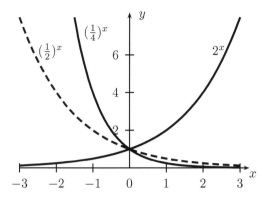

Figure 3.1: Exponential functions.

The notation for such inverse is

$$x = \log_b y, \tag{3.49}$$

where b is again the base. That is to say, x is the logarithm of y to the base b. As discussed earlier, $y = b^x > 0$ is always positive for $b > 0$, the logarithm is only possible for $y > 0$ and $b > 0$. Since $b^0 = 1$, we have $0 = \log_b 1$. From $b^1 = b$, we have $1 = \log_b b$. In addition, we can also combine the above two equations, and we have

$$b^{\log_b y} = y. \tag{3.50}$$

Since, for any real numbers p, q and $b > 0$, we can define two new variables $u = b^p, v = b^q$, and

$$y = b^p b^q = uv = b^{p+q}. \tag{3.51}$$

From the definition of the logarithm, we now have

$$p + q = \log_b y = \log_b(uv). \tag{3.52}$$

Since $p = \log_b u$ and $q = \log_b v$, we finally obtain

$$\log_b(uv) = p + q = \log_b u + \log_b v. \tag{3.53}$$

This is the multiplication rule. If we replace q by $-q$ so that $b^p b^{-q} = b^p/b^q = u/v$, we can easily verify the division rule

$$\log_b \frac{u}{v} = \log_b u - \log_b v. \tag{3.54}$$

When $p = q$ (so $u = v$), we have $\log_b(u^2) = 2\log_b u)$ which can easily be extended to any x

$$\log_b u^x = x \log_b u. \tag{3.55}$$

A special case if $x = 1/n$, we have

$$\log_b u^{1/n} = \log_b \sqrt[n]{u} = \frac{1}{n} \log_b u. \qquad (3.56)$$

As these rules are valid for any $b > 0$, it is sometimes convenient to simply write the above rule without explicitly stating the base b. This means that we more often use

$$\log(uv) = \log u + \log v. \qquad (3.57)$$

Even though any base $b > 0$ is valid, some bases are convenient for calculating logarithms and thus more widely used than others. For historical reasons and because of the decimal systems, base $b = 10$ is widely used and we write $\log_{10} u$.

Another special base is the base e for natural or Napierian logarithms where

$$e = 2.7182818284..., \qquad (3.58)$$

and in this case, we simply write the logarithm as ln, using a special notation to distinguish from the common logarithms (log).

$$\ln u \equiv \log_e u. \qquad (3.59)$$

The reasons why $e = 2.71828...$ is a natural choice for the base are many, and readers can refer to some textbooks about the history of mathematics. In this case, its corresponding exponential function

$$y = e^x, \qquad (3.60)$$

and its various forms such as $\exp[-x]$ or $\exp[-x^2]$ appear in a wide range of applications. The calculations of logarithms are not straightforward, so they were commonly listed in various mathematical tables before computers became widely used. Nowadays, even a pocket calculator can do very advanced logarithms.

3.4   Applications

3.4.1   Carbon Dioxide

The increase of the CO_2 concentration in the atmosphere will lead to an increase in the global temperature T. The simplest model is probably

$$\Delta T = \gamma \ln \frac{A(t)}{A_0}, \qquad (3.61)$$

where $A(t)$ is the CO_2 concentration at time t and A_0 is the reference concentration at $t = 0$. The parameter γ is called the climate sensitivity, and it typically takes the value of $\gamma \approx 3.6$ K. If we double the CO_2 concentration $A = 2A_0$, we have the increase in temperature $\Delta T = 3.6 \times \ln 2 \approx 2.5$ K.

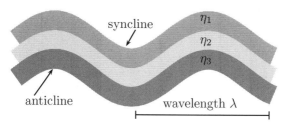

syncline

η_1

η_2

η_3

anticline

wavelength λ

Figure 3.2: Multilayered parallel folds.

3.4.2 Folds

Multilayered crust can form folds when compressed by a lateral force, resulting in buckling and deflection of the layers (see Fig. 3.2).

The dominant wavelength of folds can be described by the Biot-Ramberg equation

$$\lambda = 2\pi h \sqrt[3]{\frac{1}{6n^2} \frac{\eta_1}{\eta_2}}, \tag{3.62}$$

where h is the thickness of the layer, and λ is the wavelength of the folds. η_1/η_2 is the viscosity contrast or ratio of layers such as limestone in shale. n is the number of layers. We can see that high viscosity contrast η_1/η_2 will result in long wavelength as in the case of limestone in shale, while low viscosity ratio η_1/η_2 will typically result in short wavelength, such as siltstone and shale layers. In addition, a thicker layer (larger h) also produces folds with a longer wavelength, while multiple layers have short wavelengths compared with their monolayer counterpart.

Suppose the average viscosity ratio for a region with folds is $\eta_1/\eta_2 = 19$. What is the wavelength for a layer with a given h?

From the equation, we have $\lambda = 2\pi h(\frac{1}{6} \times 19)^{1/3} = 9.23h$. For $h = 5$ mm, we have $\lambda \approx 9.23 \times 5 \approx 46$ mm, where the symbol '$\approx$' means 'approximately equal to' or 'is about'. However, if $h = 50$ m, then $\lambda \approx 460$ m. Conversely, we can measure the wavelength of the folds and then estimate the rheological properties.

3.4.3 Gutenberg-Richter Law

The Gutenberg-Richter law is an empirical relationship between the magnitude M and the total number N of earthquakes for a given period in a given region. It can be written as

$$\log_{10} N = a - bM, \qquad \text{or} \quad N = 10^{a-bM}, \tag{3.63}$$

where a and b are constants. The value of b or b-value is of more scientific importance, and typically $0.5 \sim 2$ with a mean of $b \approx 1$.

This relationship suggests that there will be 10-fold decrease in seismic activity for a unit increase in magnitude. That is to say, there are about 10 times more magnitude 5 earthquakes than magnitude 6 earthquakes.

In general b is relatively stable; it does not vary much from region to region. But recent studies show that it may vary between individual fault zones. Whatever the variability of b will be, it is relatively safe to say that Gutenberg-Richter law will be valid statistically in most cases.

If there are N_1 earthquakes of magnitude M_1, and N_2 earthquakes of magnitude M_2, what is the ratio of N_1/N_2 if $M_2 - M_1 = 3$?

From the empirical law, we know that

$$N_1 = 10^{a-bM_1}, \qquad N_2 = 10^{a-bM_2}, \qquad (3.64)$$

which leads to

$$\frac{N_1}{N_2} = \frac{10^{a-bM_1}}{10^{a-bM_2}} = \frac{10^a 10^{-bM_1}}{10^a 10^{-bM_2}} = 10^{b(M_2-M_1)}. \qquad (3.65)$$

If $b = 1$ and $M_2 - M_1 = 3$, we have $\frac{N_1}{N_2} = 10^{1 \times 3} = 1000$. That is to say there is a factor of 1000 in terms of the numbers of earthquakes.

Another measure of the earthquake magnitude is the moment magnitude M_w which is related to the seismic moment M_0 defined by

$$M_0 = \mu A d, \qquad (3.66)$$

where μ is the rock rigidity, d is the slip distance and A is the fault area in the rupture plane. The unit of M_0 is dyne-cm (10^5 dyne $=1$ N), and all lengths are in centimetres. The moment magnitude M_w can be calculated by

$$M_w = \frac{2}{3}[\log_{10} M_0 - 16]. \qquad (3.67)$$

The seismic energy can be estimated by $E_s = M_0/20000$. For example, for the great Chilean Quake in 1960, the estimated values of slip distance $d = 30$ m $= 3 \times 10^3$ cm, and $\mu = 3 \times 10^{11}$ dyne/cm^2. The fault length is about 1000 km $= 10^8$ cm, and the width is 120 km $= 1.2 \times 10^7$ cm, which means $A = 10^8 \times 1.2 \times 10^7 = 1.2 \times 10^{15}$ cm^2. Therefore, we have

$$M_0 = 3 \times 10^{11} \times 1.2 \times 10^{15} \times 3 \times 10^3 = 1.08 \times 10^{30} \text{ dyne-cm.} \quad (3.68)$$

Its moment magnitude M_w is approximately

$$M_w = \frac{2}{3}[\log_{10}(1.08 \times 10^{30}) - 16] \approx 9.4. \qquad (3.69)$$

The seismic energy E_s is about $M_0/20000 \approx 5.4 \times 10^{25}$ ergs. Another way to estimate E_s is to use the Richter formula $\log_{10} E_s = 11.8 + 1.5 M_w \approx 11.8 + 1.5 \times 9.4 \approx 25.9$ so that $E_s = 10^{25.9} \approx 7.9 \times 10^{25}$ ergs,

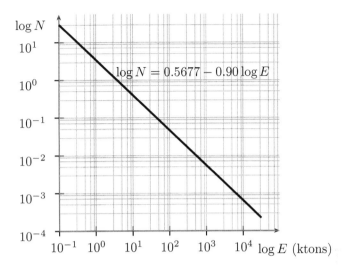

Figure 3.3: Energy-frequency distribution of objects impacting the Earth.

which is about the energy of 2 billion tons of TNT. Here we have used 1 ton of TNT which is equivalent to 4.18×10^9 J $= 4.18 \times 10^{16}$ ergs. In the calculations, we have used the approximation that the Richter-Romberg scale is about the same as the moment scale, though they are in fact different.

3.4.4 Distribution of Near-Earth Objects

According to a detailed study by the NASA's Near-Earth Object Science Definition Team, the size-frequency distribution of near-Earth asteroids (NEA) approximately obey a power-law

$$N_{>D} \approx 1148D^{-2.354}, \tag{3.70}$$

where $N_{>D}$ is the cumulative number or the number of asteroids with a diameter greater than D in the unit of km. This means there are about $1148 \times 2^{-2.354} \approx 224$ NEAs with a diameter of more than 2 km, while there are about $1148 \times (0.1)^{-2.354} \approx 260,000$ NEAs larger than 100 m.

Another study by Brown et al. (2002) suggested that the cumulative number, N, of objects colliding with the Earth each year with energies, E, in kilotons or greater (1 kiloton TNT is about 4.18×10^{12} J) obey a power-law relationship

$$\log N_{>E} = 0.5677 - 0.90 \log E, \tag{3.71}$$

or
$$N_{>E} = 10^{0.5677} E^{-0.90} \approx 3.7 E^{-0.9}.$$

For example, for $E = 5$ ktons, we have $N_{>5} \approx 3.7 \times 5^{-0.9} \approx 1$ or about once a year. For $E = 10000$ ktons $= 10$ Megatons (with the energy of Tunguska explosion), we have $N_{>10000} \approx 3.7 \times 10000^{-0.9} \approx 0.00093$ per year, or about one in every $1/0.00093 \approx 1000$ years. The log$-$log plot of equation (3.71) is shown in Fig. 3.3.

The above equation is equivalent to the following equation in terms of diameter D in metres of the colliding body

$$\log N_{>D} = 1.568 - 2.70 \log D, \tag{3.72}$$

or

$$N_{>D} = 10^{1.568} D^{-2.70} \approx 36.98 D^{-2.70}. \tag{3.73}$$

For example, if $D = 2$ m, we have $N_{>2} \approx 36.98 \times 2^{-2.7} \approx 6$ per year. This means the total number of objects with a diameter greater than 2 m colliding with the Earth is about 6 per year. If $D = 0.5$ m, we have $N_{>0.5} \approx 36.98 \times 0.5^{-2.7} \approx 240$ per year, or 20 per month. These small NEAs will usually break up and explode in the upper atmosphere without causing any noticeable damage. However, on the other hand, for $D = 100$ m, we have $N_{>100} \approx 36.98 \times 100^{-2.7} \approx 0.00015$ per year, which is equivalent to about one in every $1/0.00015 \approx 6700$ years.

Chapter 4

Trigonometry and Spherical Trigonometry

Spherical trigonometry is very important in earth sciences; however, it has many fundamental differences from its counterparts of plane trigonometry. In order to compare them, we have to review the fundamental concepts of plane trigonometry.

Spherical geometry is a natural mathematical tool for earth sciences, as the Earth's surface is approximately spherical. If our interest is regional, where the effect of the Earth's curvature is negligible, then local variations become planar. In this case, plane trigonometry becomes relevant and useful.

In this chapter, after introducing the fundamentals of trigonometry and spherical trigonometry, we will apply them to determine the variations of gravity with latitude, the great-circle distance between any two points on a sphere, the relationship between bubble pressure and surface tension, and contact angle and capillary pressure.

4.1 Plane Trigonometry

4.1.1 Trigonometrical Functions

Let us first define the angle in trigonometry. An angle is the measure in an anti-clockwise direction from the x-axis, and such an angle is positive, while a negative angle is measured in a clockwise direction from the x-axis. The angles are often expressed either in degrees $°$ or in radians. A whole circle is often divided into $360°$, and a right angle is $90°$.

A radian is defined as the angle whose arc length is r which is also the radius (see Fig. 4.1). As the total arc length or circumference of a

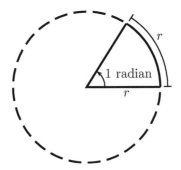

Figure 4.1: Definition of a radian.

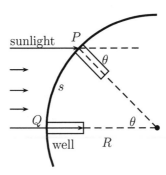

Figure 4.2: Eratosthenes' method of measuring the Earth's size.

circle is $2\pi r$, so we have

$$2\pi \text{ in radians} = 360° \text{ in degrees}, \tag{4.1}$$

which means a right angle is $\pi/2$. By dividing both sides of the above equation by 2π, we have

$$1 \text{ radian} = \frac{360°}{2 \times 3.14159} = 57.296° \text{ degrees}. \tag{4.2}$$

Conversely, we have

$$1° = \frac{2\pi}{360} = 0.017453 \text{ radians}. \tag{4.3}$$

Example 4.1: The ancient Greek geographer, Eratosthenes, was probably the first person who tried to measure the size of the Earth accurately using two vertical wells. In one well Q at Syene, the sunlight happened to

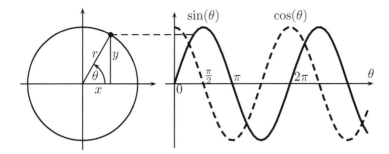

Figure 4.3: A sine function $\sin(\theta)$ (solid) and a cosine function $\cos(\theta)$ (dashed).

shine down the well at noon exactly during the Summer Solstice, while the sunlight had a angle θ with its vertical axis in another well P at Alexandria. The distance s between P and Q is known. We know the circumference L of the Earth is related to its radius R by $L = 2\pi R$. From simple trigonometry, we know that

$$\frac{s}{L} = \frac{\theta}{360°},$$

which leads to

$$L = \frac{360s}{\theta}.$$

Eratosthenes measured that $\theta \approx 7.2°$ and $s = 787$ km, so the circumference of the Earth was estimated to be

$$L \approx \frac{360 \times 790}{7.2} \approx 39,500 \text{ km},$$

which is very close to the modern value of $L = 40,040$ km.

Referring to the variables in Fig. 4.3, the basic sine and cosine functions are defined as

$$\sin\theta = \frac{y}{r}, \qquad \cos\theta = \frac{x}{r}. \tag{4.4}$$

Since $|x| \leq r$ and $|y| \leq r$ for any triangle, we have

$$-1 \leq \sin\theta \leq 1, \qquad 1 \leq \cos\theta \leq 1. \tag{4.5}$$

From the graphs of $\sin\theta$ and $\cos\theta$ in Fig. 4.3, we know that

$$\sin 0 = \sin\pi = 0, \quad \cos 0 = \cos 2\pi = 1, \quad \sin\frac{\pi}{2} = 1, \tag{4.6}$$

and

$$\cos\pi = -1, \quad \sin\frac{3\pi}{2} = -1. \tag{4.7}$$

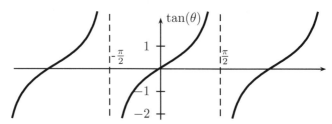

Figure 4.4: Graph of $\tan(\theta)$

In fact, the sine curve becomes a cosine curve if you shift to the left by $\pi/2$. That is

$$\sin(\theta + \frac{\pi}{2}) = \cos\theta. \tag{4.8}$$

Similarly, we have $\cos(\theta + \frac{\pi}{2}) = -\sin\theta$. If we add 2π or $360°$ to the angle θ, we will reach to the same point, we then have

$$\sin(2\pi + \theta) = \sin(\theta), \qquad \cos(2\pi + \theta) = \cos\theta. \tag{4.9}$$

This is equivalent to saying that both $\sin\theta$ and $\cos\theta$ functions have a period of 2π or $360°$.

In addition, if we replace θ by $-\theta$, then from the definition $\sin\theta = y/r$, the point at (x, y) becomes $(x, -y)$ with the same r. This means that

$$\sin(-\theta) = \frac{-y}{r} = -\sin(\theta), \tag{4.10}$$

which suggests that $\sin\theta$ is any odd function. Similar argument suggests that $\cos(-\theta) = \cos(\theta)$, which is an even function.

Other trigonometrical functions can be defined using the basic sin and cos functions. For example, the $\tan\theta$ can be defined as

$$\tan\theta = \frac{y}{x} = \frac{\sin\theta}{\cos\theta}, \tag{4.11}$$

whose graph is shown in Fig. 4.4. Similarly, we have

$$\cot\theta = \frac{1}{\tan\theta} = \frac{\cos\theta}{\sin\theta}, \qquad \operatorname{cosec}\theta = \frac{1}{\sin\theta}, \qquad \sec\theta = \frac{1}{\cos\theta}. \tag{4.12}$$

In trigonometry, we often write either $\sin(\theta)$ or $\sin\theta$ as long as there is no ambiguity. We also write the power as $\sin^n\theta \equiv (\sin\theta)^n$ for clarity and simplicity.

Example 4.2: The geodetic reference system for calculating the gravity as a function of latitude can be expressed by the Somigliana equation

$$g(\phi) = g_0 \frac{1 + k\sin^2\lambda}{\sqrt{1 - e^2\sin^2\lambda}},$$

where $g_0 = 9.7803267714$ m/s^2 is the gravity at the equator $\lambda = 0$. $k = 0.001931851386$ and $e = 0.006694379990$ are constants. This formula has been used since 1980.

Now let us calculate the value of g at the sea level for $\lambda = 45°$. Since $\sin 45° = \sqrt{2}/2$ and $\sin^2 45° = 1/2$, we have

$$g(45°) = g_0 \frac{1 + k/2}{\sqrt{1 - e^2/2}}$$

$$= 9.7803267714 \times \frac{1 + 0.001931851386/2}{\sqrt{1 - 0.006694379990^2/2}} \approx 9.8061992 \text{ m/s}^2$$

There was another international gravity formula developed in 1967

$$g(\lambda) = g_0[1 + \alpha \sin^2 \lambda - \beta \sin^2(2\lambda)],$$

where $g_0 = 9.78031846$ m/s^2, $\alpha = 5.3024 \times 10^{-3}$ and $\beta = 5.8 \times 10^{-6}$. Let us see how this formula differs from the Somigliana equation at $\lambda = 45°$. Since $\sin(90°) = 1$, we have

$$g(45°) = 9.78031846 \times (1 + 5.3024 \times 10^{-3} \times \frac{1}{2} - 5.8 \times 10^{-6} \times 1^2] \approx 9.80619.$$

Both formulae agree very well with 10^{-5} m/s^2.

- -

Using the Pythagoras' theorem $x^2 + y^2 = r^2$, we have

$$\sin^2 \theta + \cos^2 \theta = (\frac{y}{r})^2 + (\frac{x}{r})^2 = \frac{y^2 + x^2}{r^2} = \frac{r^2}{r^2} = 1. \qquad (4.13)$$

This equality is true for any angle θ, so we call it an identity. In some mathematical texts, they often use the $\equiv$ symbol to emphasise the fact that it is an identity and write it as

$$\sin^2 \theta + \cos^2 \theta \equiv 1. \qquad (4.14)$$

Again this suggests that $|\sin \theta| \leq 1$, and $|\cos \theta| \leq 1$.

Since $\tan \theta = \sin \theta / \cos \theta$ and

$$1 + \frac{\sin^2 \theta}{\cos^2 \theta} = \frac{\cos^2 \theta + \sin^2 \theta}{\cos^2 \theta} = \frac{1}{\cos^2 \theta}, \qquad (4.15)$$

we have the following identity

$$1 + \tan^2 \theta \equiv \frac{1}{\cos^2 \theta} = \sec^2 \theta. \qquad (4.16)$$

Other identities can be proved in the similar manner.

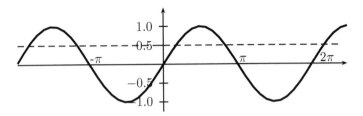

Figure 4.5: Inverse of $\sin \theta = 0.5$.

Example 4.3: Prove that the following identity

$$\frac{1}{\tan^2 \theta + 1} + \frac{1}{\cot^2 \theta + 1} \equiv 1.$$

Since $\tan \theta = \sin \theta / \cos \theta$ and $\cot \theta = \cos \theta / \sin \theta$, we have

$$\frac{1}{\tan^2 \theta + 1} + \frac{1}{\cot^2 \theta + 1} = \frac{1}{\frac{\sin^2 \theta}{\cos^2 \theta} + 1} + \frac{1}{\frac{\cos^2 \theta}{\sin^2 \theta} + 1}$$

$$= \frac{1}{\frac{\sin^2 \theta + \cos^2 \theta}{\cos^2 \theta}} + \frac{1}{\frac{\cos^2 \theta + \sin^2 \theta}{\sin^2 \theta}} = \frac{1}{\frac{1}{\cos^2 \theta}} + \frac{1}{\frac{1}{\sin^2 \theta}} = \cos^2 \theta + \sin^2 \theta = 1,$$

which becomes the original identity.

- -

The inverse of a sine function can be obtained by finding the value of θ once the value of $y = \sin \theta$ is given. That is

$$\sin \theta = y, \qquad \text{or} \qquad \theta = \sin^{-1} y. \qquad (4.17)$$

In some literature, $\sin^{-1}$ is also written as arcsin. For example, in order to find the values θ so that $\sin \theta = 1/2$, we can plot $\sin(\theta)$ for various values of θ (solid curve) and then draw a straight line shown in Fig. 4.5 where we can see that there are multiple solutions at $x = -210°, 30°,$ $150°, \ldots$ or $x = -7\pi/6, \pi/6, 5\pi/6, \ldots$. In fact, there are infinitely many solutions in the form of

$$x = \frac{\pi}{6} + 2n\pi, \qquad x = \frac{5\pi}{6} + 2n\pi, \qquad (4.18)$$

where $n = 0, \pm 1, \pm 2, \ldots$. In order to avoid any possible ambiguity, the angles in radians in the interval $[-\frac{\pi}{2}, \frac{\pi}{2}]$ are commonly defined as the principal values of the $\sin^{-1}$ function.

Similarly, the inverse of cos is denoted by $\cos^{-1}$, and its principal values are in the interval of $[0, \pi]$, while the inverse of tan is $\tan^{-1}$ whose principal values are between $-\pi/2$ and $\pi/2$, inclusive. It is worth noting

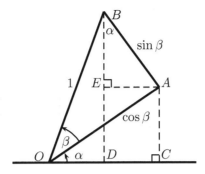

Figure 4.6: Derivation of $\sin(\alpha + \beta)$.

that, since $-1 \le \sin\theta \le 1$, the expression such as $\sin^{-1} 2$ is not valid as there exists no angle such that $\sin\theta$ is 2.

For the sine function of the addition of the two angles α and β (see Fig. 4.6), we let the length of OB be 1 or $OB = 1$. Now we have

$$\sin(\alpha + \beta) = \frac{BD}{OB} = BD = BE + ED. \tag{4.19}$$

Since $OA = \cos\beta$, $AB = \sin\beta$ and $AE = \sin\alpha\sin\beta$, we have

$$AC = ED = OA\sin\alpha = \cos\beta\sin\alpha, \quad BE = AB\cos\alpha = \sin\beta\cos\alpha,$$

which leads to

$$\sin(\alpha + \beta) = \sin\alpha\cos\beta + \cos\alpha\sin\beta. \tag{4.20}$$

In the special case when $\alpha = \beta$, we have

$$\sin 2\alpha = \sin\alpha\cos\alpha + \cos\alpha\sin\alpha = 2\sin\alpha\cos\alpha. \tag{4.21}$$

Similarly, we have

$$\cos(\alpha + \beta) = \frac{OD}{OB} = OD = OC - DC. \tag{4.22}$$

From $OC = OA\cos\alpha = \cos\beta\cos\alpha$ and $DC = EA = AB\sin\alpha = \sin\beta\sin\alpha$, we finally have

$$\cos(\alpha + \beta) = \cos\alpha\cos\beta - \sin\alpha\sin\beta. \tag{4.23}$$

Replacing β by $-\beta$ and using $\sin(-\beta) = -\sin\beta$ and $\cos(-\beta) = \cos\beta$, we have

$$\sin(\alpha - \beta) = \sin\alpha\cos\beta - \cos\alpha\sin\beta, \tag{4.24}$$

and
$$\cos(\alpha - \beta) = \cos\alpha\cos\beta + \sin\alpha\sin\beta. \qquad (4.25)$$

From equation (4.23), if $\alpha = \beta$, we have $\cos 2\alpha = \cos\alpha\cos\alpha - \sin\alpha\sin\alpha$
$= \cos^2\alpha - \sin^2\alpha$.

Since $\sin^2\alpha + \cos^2\alpha = 1$ or $\sin^2\alpha = 1 - \cos^2\alpha$, we have

$$\cos 2\alpha = \cos^2\alpha - \sin^2\alpha = 2\cos^2\alpha - 1 = 1 - 2\sin^2\alpha. \qquad (4.26)$$

Example 4.4: Prove the identity

$$\tan 2\alpha = \frac{2\tan\alpha}{1 - \tan^2\alpha}.$$

By combining (4.21) and (4.26), we have

$$\tan 2\alpha = \frac{\sin 2\alpha}{\cos 2\alpha} = \frac{2\sin\alpha\cos\alpha}{\cos^2\alpha - \sin^2\alpha}.$$

If we divide both the numerator and denominator by $\cos^2\alpha$, we have

$$\tan 2\alpha = \frac{\frac{2\sin\alpha}{\cos\alpha}}{1 - \frac{\sin^2\alpha}{\cos^2\alpha}} = \frac{2\tan\alpha}{1 - \tan^2\alpha}.$$

- -

By subtracting (4.24) from (4.20), we have

$$\sin(\alpha+\beta) - \sin(\alpha-\beta) = \sin\alpha\cos\beta + \cos\alpha\sin\beta - (\sin\alpha\cos\beta - \cos\alpha\sin\beta)$$

$$= 2\cos\alpha\sin\beta. \qquad (4.27)$$

Similarly, by adding the two identities (4.20) and (4.24), we have

$$\sin(\alpha + \beta) + \sin(\alpha - \beta) = 2\sin\alpha\cos\beta. \qquad (4.28)$$

By introducing new notations $A = \alpha + \beta$, and $B = \alpha - \beta$, we have after simple addition and substraction

$$\alpha = \frac{A+B}{2}, \qquad \beta = \frac{A-B}{2}. \qquad (4.29)$$

The above identities (4.27) and (4.28) now become

$$\sin A - \sin B = 2\cos\frac{A+B}{2}\sin\frac{A-B}{2}, \qquad (4.30)$$

and

$$\sin A + \sin B = 2\sin\frac{A+B}{2}\cos\frac{A-B}{2}. \qquad (4.31)$$

The formula for the addition of two angles can be used to derive a general expression

$$A \cos \theta + B \sin \theta = R \cos(\theta - \psi), \tag{4.32}$$

for any θ and constants A and B. From (4.25), we know that

$$\cos(\theta - \psi) = \cos \theta \cos \psi + \sin \theta \sin \psi. \tag{4.33}$$

Multiplying both sides by R, we have

$$R \cos(\theta - \psi) = (R \cos \psi) \cos \theta + (R \sin \psi) \sin \theta. \tag{4.34}$$

By comparing with the original equation (4.32), we have

$$A \cos \theta + B \sin \theta = (R \cos \psi) \cos \theta + (R \sin \psi) \sin \theta. \tag{4.35}$$

This must be true for any θ, so we have

$$A = R \cos \psi, \qquad B = R \sin \psi. \tag{4.36}$$

Taking squares and using $\cos^2 \psi + \sin^2 \psi = 1$, we have

$$A^2 + B^2 = R^2 \cos^2 \psi + R^2 \sin^2 \psi = R^2(\cos^2 \psi + \sin^2 \psi) = R^2, \tag{4.37}$$

which gives

$$R = \sqrt{A^2 + B^2}. \tag{4.38}$$

Here we have only used the positive root as R is conventionally considered as the amplitude of the variations.

By taking the ratio of B and A, we have

$$\frac{B}{A} = \frac{R \sin \psi}{R \cos \psi} = \tan \psi, \tag{4.39}$$

or

$$\psi = \tan^{-1} \frac{B}{A}. \tag{4.40}$$

4.1.2 Sine Rule

The sine rule is important in trigonometry as it provides the relationships between the three sides and their corresponding angles. The height h in the triangle shown in Fig. 4.7 can be calculated using

$$h = b \sin A, \qquad \text{or } h = a \sin B, \tag{4.41}$$

depending which right-angled smaller triangle you are using. The combination of the above two formulae leads to

$$b \sin A = a \sin B, \quad \text{or} \quad \frac{a}{\sin A} = \frac{b}{\sin B}. \tag{4.42}$$

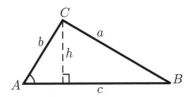

Figure 4.7: Derivation of the sine rule.

Applying the same argument to other perpendicular heights, we can obtain

$$\frac{b}{\sin B} = \frac{c}{\sin C}. \tag{4.43}$$

Combing the above equations, we have the sine rule

$$\frac{a}{\sin A} = \frac{b}{\sin B} = \frac{c}{\sin C}. \tag{4.44}$$

This rule makes it easier to determine all other two sides very easily once the length of a side and two angles are given in a triangle.

On the other hand, if one angle and its two sides are given, it is easier to use the cosine rule to determine the rest quantities of the given triangle.

4.1.3   Cosine Rule

In order to derive the cosine rule, we refer to the same triangle in deriving the sine rule, but now we put it in a Cartesian coordinate system shown in Fig. 4.8. Now the coordinates are $(0,0), (c,0)$ and $(b \cos A, b \sin A)$. The length a is simply the Cartesian distance between points A and B. We now have

$$a^2 = (b \cos A - c)^2 + (b \sin A - 0)^2 = b^2 \cos^2 A + c^2 - 2bc \cos A + b^2 \sin^2 A$$

$$= b^2(\cos^2 A + \sin^2 A) + c^2 - 2bc \cos A. \tag{4.45}$$

By using $\sin^2 A + \cos^2 A = 1$, we finally obtain the cosine rule

$$a^2 = b^2 + c^2 - 2bc \cos A. \tag{4.46}$$

Similarly, by proper permutation using the same argument, we have

$$b^2 = c^2 + a^2 - 2ca \cos B, \qquad c^2 = a^2 + b^2 - 2ab \cos C. \tag{4.47}$$

By rearranging the cosine rule, we can obtain the formulae to determine the angles of a triangle once the lengths of the three sides are given. We have

$$\cos A = \frac{b^2 + c^2 - a^2}{2bc}, \qquad \cos B = \frac{c^2 + a^2 - b^2}{2ca}, \tag{4.48}$$

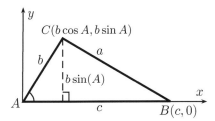

Figure 4.8: Derivation of the cosine rule.

and

$$\cos C = \frac{a^2 + b^2 - c^2}{2ab}. \tag{4.49}$$

Example 4.5:   In the triangle shown in Fig. 4.7, we have the angle $A = 60°$, and two sides $b = 2$ and $c = 4$. Now we try to determine the other two angles and a. First, using the cosine rule, we have

$$a^2 = b^2 + c^2 - 2bc \cos A = 2^2 + 4^2 - 2 \times 2 \times 4 \cos 60° = 4 + 16 - 16\frac{1}{2} = 12,$$

or $a = \sqrt{12}$. In order to determine the angle B, we can use (4.48)

$$\cos B = \frac{c^2 + a^2 - b^2}{2ca} = \frac{4^2 + (\sqrt{12})^2 - 2^2}{2 \times 4 \times \sqrt{12}} = \frac{3}{2\sqrt{3}} = \frac{\sqrt{3}}{2},$$

which gives

$$B = \cos^{-1} \frac{\sqrt{3}}{2} = 30°.$$

Thus, the other angle is $C = 180° - A - B = 90°$. This means that it is a right-angled triangle.

4.2   Spherical Trigonometry

Spherical trigonometry is more commonly used in earth sciences and astronomy. Compared to plane trigonometry, spherical trigonometry is more complicated. Here we will give a brief introduction and highlight the most important concepts without providing proof. Interested readers can refer to more advanced books on spherical geometry.

In plane geometry, we often think that between any two different points there exists a straight line which is the shortest; however, in spherical geometry, the concept of lines has to be changed. The shortest distance connecting any two points on a sphere is a portion of a great

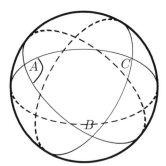

Figure 4.9: Great circles and spherical triangles.

circle, where a great circle is a circle on the sphere such that it divides the sphere into two equal parts in area, and the centre of the great circle is always the centre of the sphere.

For example, if we consider the Earth as a perfect sphere, then the lines of longitude and the equator are great circles. On a plane, two parallel lines will not meet, while on a sphere, any two lines will meet at two points.

A spherical triangle is formed by three great circles (see Fig. 4.9), and its three corresponding angles $\angle A$, $\angle B$ and $\angle C$ on the sphere are the angles formed by these great circles. These angles are dihedral face angles. In plane trigonometry, the sum of all the three internal angles of a triangle is always $180°$ or π; however, in spherical trigonometry the sum of the three angles of a spherical triangle is always greater than $180°$. In fact, the area S of a spherical triangle $\triangle ABC$ can be calculated using

$$S = \frac{\pi R^2}{180°}(A + B + C - 180°), \tag{4.50}$$

where R is the radius of the sphere, and the angles are in degrees.

In plane geometry, a triangle is a polygon with the smallest number of sides. There is no two-sided closed shape. But on a sphere, a lune is a two-sided shape formed by two great circles. As the total surface area of a sphere is $4\pi R^2$, it is convenient to define the solid spherical angle of a sphere as 4π or $720°$.

For a spherical triangle, there are other important angles, called trihedral angles α, β, and γ formed by three half lines radiating from the centre O of the sphere (see Fig. 4.10). Since the radius R of the sphere is fixed, the length or distance of the edges of a spherical triangle are often expressed in terms of the trihedral angles, and the real length can be calculated by multiplying by R. The sine rule for a spherical

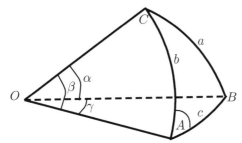

Figure 4.10: A spherical triangle and its angles.

triangle can be written as

$$\frac{\sin a}{\sin A} = \frac{\sin b}{\sin B} = \frac{\sin c}{\sin C}. \tag{4.51}$$

The first law of cosines for trihedrals is

$$\cos\alpha = \cos\beta\cos\gamma + \sin\beta\sin\gamma\cos A. \tag{4.52}$$

Other formulae for other angles can be written by proper permutation of the angles. Sometimes, the first law of cosines is also written in terms of the three sides

$$\cos a = \cos b\cos c + \sin\beta\sin c\cos A. \tag{4.53}$$

The second law of cosines is given by

$$\cos A = -\cos B\cos C + \sin B\sin C\cos\alpha, \tag{4.54}$$

and others can be written in similar forms. When the angle $C = 90°$, the triangle becomes a right spherical triangle. Since $\sin 90° = 1$ and $\cos 90° = 0$, the above second law of cosines becomes $\cos A = -\cos B \times 0 + \sin B \times 1 \times \cos\alpha = \sin B\cos\alpha$.

4.3 Applications

4.3.1 Travel-Time Curves of Seismic Waves

Similar to light, seismic waves obey Snell's law when refracted at the interface of two layers with different velocities and densities. This is true for both P-waves and S-waves. However, things are slightly more complicated as there is a mode conversion for seismic waves. That is, an incident P-wave will generate both P- and S-waves at the interface, so is the case for an incident S-wave. The angle of reflection of the same wave type will be equal to the angle of incidence. So for an incident

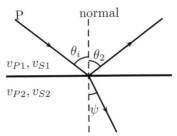

Figure 4.11: Snell's law for seismic waves at the interface of two different layers.

P-wave, its angle of reflection $\theta_{2P} = \theta_i$ (see Fig. 4.11). However, for the generated S-wave, the angle of reflection θ_2 is related to θ_i by

$$\frac{\sin\theta_i}{v_{P1}} = \frac{\sin\theta_{2S}}{v_{S1}}. \tag{4.55}$$

Since P-waves travel faster than S-waves ($v_{P1} > v_{S1}$), the angle of reflection of an S-wave is smaller than that of a P-wave. That is $\theta_{2S} \leq \theta_{2P} = \theta_i$.

The angle of refraction ψ_P for the P-wave and the angle of refraction ψ_S for the S-wave can be related to θ_i by Snell's law

$$\frac{\sin\theta_i}{\sin\psi_P} = \frac{v_{P1}}{v_{P2}}, \qquad \frac{\sin\theta_i}{\sin\psi_S} = \frac{v_{P1}}{v_{S2}}. \tag{4.56}$$

Since the speeds of P-waves and S-waves are different, P-waves and S-waves will in general have different angles of refraction.

If the angle of incidence (θ) increases, the angle of refraction (ψ) will also increase. The critical angle θ_c occurs when $\psi = 90°$. In this case, $\sin\psi = 1$, which means the refraction wave will travel along the interface. For a P-wave, we now have

$$\sin\theta_c = \frac{v_{P1}}{v_{P2}}, \tag{4.57}$$

or

$$\theta_c = \sin^{-1}\left(\frac{v_{P1}}{v_{P2}}\right). \tag{4.58}$$

This is possible only if $v_{P1} < v_{P2}$. The critical angle for an S-wave can be defined in a similar manner.

For simplicity in the rest of our discussion, we will focus on P-waves only. The simplest setting for seismic P-waves reflected in a layer over a half-space is shown in Fig. 4.12. The thickness of the layer is h with

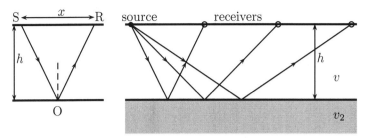

Figure 4.12: Offset and moveout of reflected P-waves.

a constant P-wave speed of v. The reflection time t_0 is the time taken for the reflection with the normal incidence $\theta_i = 0$. We have

$$t_0 = \frac{2h}{v}. \tag{4.59}$$

The distance between the source (S) and a receiver (R) is called the offset x. In addition, the direct travel time for a P-wave travelling from S along the surface to R is simply

$$t_d = \frac{x}{v}. \tag{4.60}$$

The reflection time $T(x)$ is the time taken to travel from S to O and from O to R. The total distance is d =SO+OR= 2OR. Using Pythagoras' theorem, we have OR= $\sqrt{h^2 + (x/2)^2}$. Now we have

$$T = \frac{d}{v} = \frac{2\sqrt{h^2 + x^2/4}}{v} = \frac{\sqrt{4h^2 + x^2}}{v}. \tag{4.61}$$

Taking the square of both sides and using $2h = vt_0$, we have

$$T^2 = t_0^2 + \frac{x^2}{v^2}. \tag{4.62}$$

Defining the normal moveout $\Delta t = T - t_0$, we have

$$\Delta t = T - t_0 = \sqrt{t_0^2 + \frac{x^2}{v^2}} - t_0, \tag{4.63}$$

which is essentially a hyperbola of moveout versus offset as shown in Fig. 4.13.

4.3.2 Great-Circle Distance

As an application example, now let us try to calculate the distance d between Cambridge University (Cambridge, England) and MIT (Cambridge,USA). We know that Cambridge, England is at P(52.21° N,

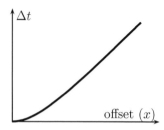

Figure 4.13: A hyperbola of moveout against offset.

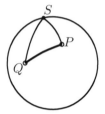

Figure 4.14: Distance between two cities.

0.12°W), and Cambridge, USA is at Q(42.37° N,71.11° W) schematically shown in Fig. 4.14) with a point S at the north pole so that PQS forms a spherical triangle by three great circles. Now we assume that the Earth is a perfect sphere. The distance or angle of SP is $SP = 90° - 52.21° = 37.79°$. Similarly, we have $SQ = 90° - 42.37 = 47.63°$, and the angle

$$\alpha = \angle QSP = 71.11° - 0.12° = 70.99°. \qquad (4.64)$$

Using the cosine rule, we have

$$\cos PQ = \cos SQ \cos SP + \sin SQ \sin SP \cos \alpha$$

$$= \cos 47.63° \cos 37.79° + \sin 47.63° \sin 37.79° \cos 70.99° = 0.6800,$$

which gives $PQ = \cos^{-1} 0.6800 = 47.15°$. Using the radius of $R = 6400$ km, we have the distance

$$d = R \times \frac{PQ\pi}{180} \approx \frac{6400 \times 47.15 \times 3.14159}{180} \approx 5267 \text{ km}, \qquad (4.65)$$

which is about 3290 miles.

4.3.3   Bubbles and Surface Tension

Soap bubbles and many relevant fantastic phenomena exist due to surface tension. Bubbles are spherical because a sphere is the geometry

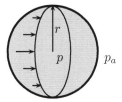

Figure 4.15: Bubble pressure and surface tension.

with a minimum energy at equilibrium. For a given liquid with the interfacial or surface tension γ, we can estimate the pressure difference between the pressure p inside a bubble with a radius r and the ambient pressure p_a outside the bubble (see Fig. 4.15). Imagine that we consider only half of the bubble; the total force is the product of the pressure difference $p - p_a$ and the cross section area πr^2, that is $F = (p - p_a)\pi r^2$. This force must be balanced by the force due to surface tension, that is $\tau = 2\gamma L$ where $L = 2\pi r$ is the circumference of the circle, and the factor 2 comes from the fact that there are two surfaces or interfaces (inside and outside). Therefore, we have $F = \tau$, or

$$(p - p_a)\pi r^2 = 2\gamma(2\pi r), \qquad (4.66)$$

which leads to

$$p - p_a = \frac{4\gamma}{r}. \qquad (4.67)$$

In the case of a water droplet (such as a dew drop) and a bubble inside liquid magma, there is only one interface, and the above formula becomes

$$p - p_a = \frac{2\gamma}{r}. \qquad (4.68)$$

This pressure difference is also called Lagrange pressure.

For example, the surface tension of water is $\gamma = 72 \times 10^{-3}$ N/m at room temperature. A bubble of size $r = 0.1$ mm $= 0.1 \times 10^{-3}$ m will have a pressure difference

$$p - p_a = \frac{4 \times 72 \times 10^{-3}}{0.1 \times 10^{-3}} = 1440 \text{ Pa} \approx 0.0144 \text{ atm}, \qquad (4.69)$$

where we have used 1 atm $= 1.01325 \times 10^5$ Pa.

Surface tension is also responsible for a phenomenon called capillary action, which is the adhesion of the liquid to the walls. The height h of the capillary action is related to surface tension γ (see Fig. 4.16). The whole weight of the liquid column inside a cylindrical tube is supported by the action of adhesion force due to surface tension. The weight of the column is $W = \rho g V$ where $V = \pi r^2 h$ is the volume, while the

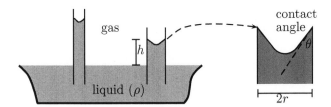

Figure 4.16: Capillary pressure and contact angle.

adhesion force is $F_a = L\gamma$ is the product of the circumference (along the wall) $L = 2\pi r$ and the surface tension γ (force per unit length). At equilibrium, we have $F_a = W$ or $2\pi r\gamma = \rho g(\pi r^2 h)$, leading to

$$h = \frac{2\gamma}{\rho g r}. \tag{4.70}$$

For example, for a tube with a radius $R = 0.1$ mm $= 10^{-4}$ m, the height of the water column due to the capillary action can be estimated. Using the typical values of $\gamma = 72 \times 10^{-3}$ N/m, $\rho = 1000$ kg/m^3, and $g = 9.8$ m/s^2, we have $h = \frac{2 \times 72 \times 10^{-3}}{1000 \times 9.8 \times 10^{-4}} \approx 0.15$ m or 15 cm high.

Since the pressure is $p = \rho g h$, the above equation can also be expressed as the capillary pressure p_c

$$p_c = \rho g h = \frac{2\gamma}{r}. \tag{4.71}$$

This is similar to equation (4.68).

The tension between liquid and gas, called interfacial tension, will usually result in a curved surface inside the tube so that a contact angle θ is formed between the tangent of the liquid surface and the solid wall. If we use detailed and rather complicated mechanical analysis, the capillary pressure should be modified slightly as $p_c = 2\gamma \cos\theta/r$.

Another important link is the relationship between the capillary pressure and the degree of saturation in porous media. In this case, the capillary pressure p_c is the difference of the pressure between non-wetting phase p_n and the pressure of wetting phase p_w, that is $p_c = p_n - p_w$ which is a monotonically decreasing function $f(S)$ of the effective water saturation S. For example, the van Genuchten empirical relationship can be expressed as

$$f(S) = p_0(S^{-1/\lambda} - 1)^{1/n}, \tag{4.72}$$

where p_0 is a constant, and $\lambda, n > 0$ are two parameters that are related to the pore size distribution. Typically $\lambda = 1 \sim 4$ and $n = 0.9 \sim 10$. The actual relationship is far more complicated with hysteresis.

Chapter 5

Complex Numbers

In the real world, all quantities are real numbers. However, sometimes it is more convenient to do mathematical analysis using complex numbers, especially for solving differential equations. In addition, we can also find the interesting links between the trigonometrical functions and hyperbolic functions using complex numbers. We will explain such links in detail in this chapter. Furthermore, we will apply hyperbolic functions to study the variations of wave velocity under various conditions, which can be applied to tsunami, shallow water waves and deep water waves.

5.1 Complex Numbers

When we discuss the square root in quadratic equations, we say the square root of a negative number, for example $\sqrt{-1}$, does not exist because there is no number whose square is negative (In fact, for all real numbers x, $x^2 \geq 0$). Well, this is only true in the context of real numbers. Such limitations mean that the system of real numbers is incomplete, as the mathematical operations of numbers, including $\sqrt{\ }$, could lead to something which does not belong to the number system.

5.1.1 Imaginary Numbers

A significant extension is to introduce imaginary numbers by defining an imaginary unit

$$i = \sqrt{-1}, \qquad i^2 = (\sqrt{-1})^2 = -1. \tag{5.1}$$

This is a seemingly simple step but it may have many profound consequences. It is worth pointing out that i is a special notation, so you cannot use $i^2 = (\sqrt{-1})^2 = \sqrt{(-1)^2} = \sqrt{1} = \pm 1$ because this may lead to some confusion. To avoid such possible confusion, it is better to think of i as $\sqrt{-}$ or the imaginary unit, so for any real number $a > 0$

$$\sqrt{-a} = \sqrt{(-1) \times a} = \sqrt{-1}\sqrt{a} = i\sqrt{a}. \tag{5.2}$$

For example, $\sqrt{-2} = i\sqrt{2}$ and $\sqrt{-25} = i5 = 5i$ (often we prefer to write numbers first followed by i).

Example 5.1:   The imaginary number i follows the same rules for mathematical functions defined in the real-number system. For example, we can calculate the following

$$i^3 = i^2 i = -1i = -i, \quad i^4 = (i^2)^2 = (-1)^2 = 1, \qquad i^5 = i^4 i = i,$$

and thus

$$i^{50} = i^{4 \times 12 + 2} = (i^4)^{12} \times i^2 = -1,$$

and

$$i^{101} = i^{4 \times 25 + 1} = i.$$

So the best way to estimate such an expression is to try to write exponents in terms of the multiple of 4 (as close as possible). To simply use odd or even exponents is not enough, and often leads to incorrect results.

You might ask what $\sqrt{i}$ or $i^{1/2}$ is? Of course, we can calculate it, but we have to define the algebraic rule for complex numbers. Before we proceed to do this, let us try to represent complex numbers geometrically.

Another way of thinking of a complex number is to represent it as a pair of two numbers x and y on a complex plane (see Fig. 5.1), similar to the Cartesian coordinates (x, y) where each point corresponds uniquely to an ordered pair of two coordinates (real numbers), but with a significant difference. Here the vertical axis is no longer a real axis; it becomes the imaginary axis iy-axis. Similarly, a point on the complex plane corresponds also uniquely to a complex number

$$z = a + bi, \tag{5.3}$$

which consists of a real part a and an imaginary part bi. This also means that it corresponds to an ordered pair of real numbers a and b. For a given $z = a + bi$, the real part is denoted by $a = \Re(z)$ and imaginary part is denoted by $b = \Im(z)$.

Now we can define the mathematical operations of the two complex numbers $z_1 = a + bi$ and $z_2 = c + di$. The addition of two complex numbers is carried out by adding their real parts and imaginary parts respectively. That is

$$(a + bi) + (c + di) = (a + c) + (b + d)i. \tag{5.4}$$

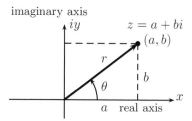

Figure 5.1: Complex plane (or Argand diagram), argument or angle and modulus.

Similarly, the subtraction is defined by

$$(a + bi) - (c + di) = (a - c) + (b - d)i. \qquad (5.5)$$

Two complex numbers are equal if, and only if, their real parts and imaginary parts are equal, respectively. That is, $z_1 = z_2$ if, and only if, $a = c$ and $b = d$.

The multiplication of two complex numbers is carried out similarly to expanding an expression, using $i^2 = -1$ when necessary

$$(a + bi) \cdot (c + di) = a \cdot (c + di) + bi \cdot (c + di)$$

$$= ac + adi + bci + bdi^2 = (ac - bd) + (bc + ad)i. \qquad (5.6)$$

The division of two complex numbers

$$\frac{a + bi}{c + di} = \frac{(a + bi) \cdot (c - di)}{(c + di) \cdot (c - di)}$$

$$= \frac{ac - adi + bci - bdi^2}{c^2 - d^2 i^2} = \frac{(ac + bd)}{(c^2 + d^2)} + \frac{(bc - ad)}{(c^2 + d^2)}i, \qquad (5.7)$$

where we have used the equality $(x + y)(x - y) = x^2 - y^2$.

Example 5.2: Now let us try to find $\sqrt{i}$. Let $z = a + bi$ and $z^2 = i$, we have

$$(a + bi)^2 = (a - b) + 2abi = i = 0 + i.$$

As two complex numbers, here $(a - b) + 2abi$ and $0 + i$, must be equal on both sides, we have

$$a - b = 0, \qquad 2ab = 1.$$

The first condition gives $a = b$, and the second gives

$$a = b = \frac{1}{\sqrt{2}} = \frac{\sqrt{2}}{2}.$$

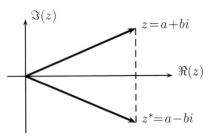

Figure 5.2: Complex conjugate and reflection.

Therefore, we have

$$\sqrt{i} = \frac{\sqrt{2}}{2} + \frac{\sqrt{2}}{2}i.$$

- -

The complex conjugate z^* of a complex number $z = a+bi$ is defined by changing the sign of the imaginary part

$$z^* = a - bi, \tag{5.8}$$

which is the reflection in the real axis of the original z (see Fig. 5.2). The definition also leads to

$$(z^*)^* = (a - bi)^* = a + bi = z. \tag{5.9}$$

5.1.2   Polar Form

From Fig. 5.1, we can also represent a complex number as a polar form in terms of an ordered pair (r, θ). From basic trigonometry, we know that

$$\sin(\theta) = \frac{b}{r}, \qquad \cos(\theta) = \frac{a}{r}, \tag{5.10}$$

or

$$a = r\cos(\theta), \qquad b = r\sin(\theta), \tag{5.11}$$

where θ is the argument or the angle of z, and r is the modulus or magnitude of the complex number $z = a + bi$, which can be obtained using Pythagoras' theorem

$$r = |z| = \sqrt{a^2 + b^2}. \tag{5.12}$$

For any given $z = a + bi$, the angle θ is given by

$$\theta = \arg(z) = \tan^{-1}\left(\frac{b}{a}\right), \tag{5.13}$$

where we only use the range $-\pi < \theta \leq \pi$, called the principal values. Thus, the same complex number can be expressed as the polar form

$$z = a + bi = r\cos(\theta) + ir\sin(\theta). \tag{5.14}$$

Sometimes, it is also conveniently written as the exponential form

$$z = re^{i\theta} = r[\cos(\theta) + i\sin(\theta)], \tag{5.15}$$

which requires the Euler formula

$$e^{i\theta} = \cos(\theta) + i\sin(\theta). \tag{5.16}$$

The proof of this formula usually involves the expansion of infinite power series, and is beyond the scope of this book.

An interesting extension is to replace θ by $-\theta$ in Euler's formula; we have

$$e^{-i\theta} = \cos(-\theta) + i\sin(-\theta) = \cos(\theta) - i\sin(\theta). \tag{5.17}$$

Adding this to the original formula (5.16), we have

$$e^{i\theta} + e^{-i\theta} = [\cos(\theta) + \cos(\theta)] + i[\sin(\theta) - \sin(\theta)] = 2\cos(\theta), \tag{5.18}$$

or

$$\cos(\theta) = \frac{e^{i\theta} + e^{-i\theta}}{2}. \tag{5.19}$$

If we follow the same procedure, but subtract these two formulae, we have

$$e^{i\theta} - e^{-i\theta} = [\cos(\theta) - \cos(\theta)] + 2\sin(\theta)i, \quad \text{or} \quad \sin(\theta) = \frac{e^{i\theta} - e^{-i\theta}}{2i}. \tag{5.20}$$

The polar form is especially convenient for multiplication, division, exponential manipulations and other mathematical manipulations. For example, the complex conjugate $z = re^{i\theta}$ is simply $z^* = re^{-i\theta}$.

Example 5.3: For example, for two complex numbers $z_1 = r_1 e^{i\theta_1}$ and $z_2 = r_2 e^{i\theta_2}$, their product is simply

$$z_1 z_2 = r_1 e^{i\theta_1} \times r_2 e^{i\theta_2} = r_1 r_2 e^{i(\theta_1 + \theta_2)}.$$

Their ratio is

$$\frac{z_1}{z_2} = \frac{r_1 e^{i\theta_1}}{r_2 e^{i\theta_2}} = \frac{r_1}{r_2} e^{i(\theta_1 - \theta_2)}.$$

Furthermore, for $z = re^{i\theta}$, we have

$$z^n = (re^{i\theta})^n = r^n (e^{i\theta})^n = r^n [\cos(\theta) + i\sin(\theta)]^n.$$

Also using Euler's formula, we have

$$z^n = r^n(e^{i\theta})^n = r^n e^{in\theta} = r^n[\cos(n\theta) + i\sin(n\theta)].$$

Combining the above two equations, we have

$$[\cos(\theta) + i\sin(\theta)]^n = \cos(n\theta) + i\sin(n\theta),$$

which is the famous de Moivre's formula.

If we now revise the previous example, we have $n = 1/2$

$$\sqrt{i} = i^{1/2} = [\cos(\frac{\pi}{2}) + i\sin(\frac{\pi}{2})]^{1/2} = \cos\frac{\pi}{4} + i\sin(\frac{\pi}{4}) = \frac{\sqrt{2}}{2} + \frac{\sqrt{2}}{2}i.$$

- -

The real functions such as logarithms can be extended directly to the functions of complex numbers. For example, we know from Euler's formula when $\theta = \pi$

$$e^{i\pi} = \cos(\pi) + i\sin(\pi) = -1 + 0i = -1. \tag{5.21}$$

If we use the logarithm definition, we should have

$$i\pi = \log_e(-1) = \ln(-1). \tag{5.22}$$

When we discussed $\log(x)$, we assumed that $x > 0$. Now we can see that $x < 0$ is also valid, though its logarithm is a complex number.

5.2 Hyperbolic Functions

Hyperbolic functions have many similarities with trigonometrical functions, and are commonly used in mechanics and earth sciences.

5.2.1 Hyperbolic Sine and Cosine

The hyperbolic sine and cosine functions are defined as

$$\sinh x = \frac{e^x - e^{-x}}{2}, \qquad \cosh x = \frac{e^x + e^{-x}}{2}, \tag{5.23}$$

where x is a real number. If we replace x by $-x$, and substitute it into the above definitions, we have

$$\sinh(-x) = \frac{e^{-x} - e^{-(-x)}}{2} = \frac{-[e^x - e^{-x}]}{2} = -\sinh x, \tag{5.24}$$

and

$$\cosh(-x) = \frac{e^{-x} + e^{-(-x)}}{2} = \frac{e^{-x} + e^x}{2} = \cosh x. \tag{5.25}$$

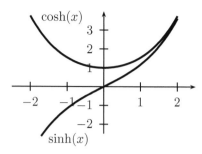

Figure 5.3: Graphs of $\sinh(x)$ and $\cosh(x)$.

This suggests that $\sinh x$ is an odd function while $\cosh x$ is an even function. Their graphs are shown in Fig. 5.3.

If we replace x by ix in the hyperbolic cosine function, we know from (5.19) in our earlier discussion that

$$\cosh(ix) = \frac{e^{ix} + e^{-ix}}{2} = \cos x, \quad \sinh(ix) = \frac{e^{ix} - e^{-ix}}{2} = i \sin x, \quad (5.26)$$

where we have used $\sin x = (e^{ix} - e^{-ix})/2$ from (5.20). Other hyperbolic functions are defined in a similar manner to the ratio of the basic hyperbolic sine and cosine functions. For example, we have the hyperbolic tangent $\tanh x = \frac{\sinh x}{\cosh x} = \frac{e^x - e^{-x}}{e^x + e^{-x}}$, and $\coth x = 1/\tanh x$, sech $x = 1/\cosh x$, and cosech $x = 1/\sinh x$.

5.2.2 Hyperbolic Identities

The hyperbolic functions also have similar identities to their trigonometrical counterparts.

Example 5.4: In order to prove $\cosh^2 x - \sinh^2 x = 1$, we start from

$$\cosh^2 x - \sinh^2 = (\cosh x + \sinh x)(\cosh x - \sinh x)$$

$$= (\frac{e^x + e^{-x}}{2} + \frac{e^x - e^{-x}}{2})(\frac{e^x + e^{-x}}{2} - \frac{e^x - e^{-x}}{2}) = (\frac{2e^x}{2})(\frac{2e^{-x}}{2}) = 1.$$

This is similar to $\cos^2 x + \sin^2 x = 1$.

- -

There is a quick way to obtain the corresponding identities from those identities for trigonometrical functions. Taking the squares of both sides of (5.26), we have

$$\cos^2 x = \cosh^2(ix), \qquad \sin^2 x = \frac{1}{i^2} \sinh^2(ix) = -\sinh^2(ix). \quad (5.27)$$

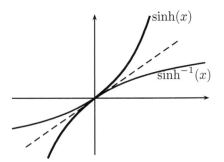

Figure 5.4: Graphs of $\sinh(x)$ and $\sinh^{-1}(x)$.

This implies that we can replace $\cos^2 x$ by $\cosh^2 x$ and $\sin^2 x$ by $-\sinh^2 x$ in the identity $\cos^2 x + \sin^2 x = 1$; we will get the identity $\cosh^2 x - \sinh^2 x = 1$ for hyperbolic functions. That is Osborn's rule for converting identities where we only need to change the sign of any terms containing the squares of a sine function including tangent (tan) and cotangent (cot). For example, we can change

$$\tan 2\theta = \frac{2\tan\theta}{1 - \tan^2\theta}, \quad \text{into} \quad \tanh 2x = \frac{2\tanh x}{1 + \tanh^2 x}. \tag{5.28}$$

In fact, we can even extend Osborn's rule further to include $\cos \rightarrow \cosh$ and $\sin \rightarrow i\sinh$. For example, from the identity $\cos(\alpha + \beta) = \cos\alpha\cos\beta - \sin\alpha\sin\beta$, we have

$$\cosh(\alpha + \beta) = \cosh\alpha\cosh\beta - i\sinh\alpha \times i\sinh\beta$$

$$= \cosh\alpha\cosh\beta + \sinh\alpha\sinh\beta, \tag{5.29}$$

where we have used $i^2 = -1$.

5.2.3   Inverse Hyperbolic Functions

The inverse of $\sinh x$ can easily be obtained (graphically shown in Fig. 5.4) by simply reflecting the graph of $\sinh x$ in the line of $y = x$ (dashed line). Mathematically speaking, we want to find $y = \sinh^{-1} x$ such that $\sinh y = x$. From the identity $\cosh^2 y - \sinh^2 y = 1$, we have

$$\cosh^2 y = 1 + \sinh^2 y = 1 + x^2. \tag{5.30}$$

Since $\cosh y = (e^y + e^{-y})/2 \geq 1$ (see Fig. 5.3), the above equation becomes

$$\cosh y = \sqrt{1 + x^2}. \tag{5.31}$$

In addition, from the previous example, we know that $\cosh y + \sinh y = \frac{e^y + e^{-y}}{2} + \frac{e^y - e^{-y}}{2} = e^y$. After combining the above two equations, we now have

$$\sqrt{1 + x^2} + \sinh y = e^y. \tag{5.32}$$

Using $\sinh y = x$, we have $\sqrt{1 + x^2} + x = e^y$. Taking the logarithm of both sides, we finally obtain

$$y = \sinh^{-1} x = \ln[x + \sqrt{1 + x^2}]. \tag{5.33}$$

The inverse of other functions can be obtained in a similar manner.

Example 5.5: In order to get $y = \cosh^{-1} x$ or $\cosh y = x$, we use

$$\cosh^2 y - \sinh^2 y = 1,$$

so that $\sinh^2 y = \cosh^2 y - 1 = x^2 - 1$, which gives $\sinh y = \sqrt{x^2 - 1}$ where $(x \geq 1)$. Again using the identity $\cosh y + \sinh y = e^y$, we have

$$x + \sinh y = x + \sqrt{x^2 - 1} = e^y.$$

Taking the logarithms, we now have

$$y = \cosh^{-1} x = \ln[x + \sqrt{x^2 - 1}]$$

where $x \geq 1$. It is worth pointing out that there are two branches of $\cosh^{-1} x$, and we have assumed here that $\cosh^{-1} x > 0$.

Complex numbers have a wide range of applications as many mathematical techniques become simpler and more powerful using complex numbers. For example, in the discrete Fourier series to be introduced later in this book, the formulae are simpler when written in the form of complex numbers.

5.3 Water Waves

Water waves in oceans and lakes are linked to many processes and phenomena. In most cases, wave forms can conveniently be expressed as

$$\psi(x, t) = A \sin(kx - \omega t), \qquad \text{or} \quad \psi(x, t) = A \cos(kx - \omega t), \tag{5.34}$$

where $\omega = 2\pi/T$ is the angular frequency of the wave, T is its period, and A is its amplitude. x is the distance and t is time. Here $k = 2\pi/\lambda$ is the wavenumber where λ is the wavelength, and it essentially means the numbers of waves in a unit length. In this case, the wave speed or phase speed v is $v = \omega/k$.

For water waves, the phase speed or velocity v in most cases is governed by

$$v = \sqrt{\frac{g\lambda}{2\pi} \tanh(\frac{2\pi h}{\lambda})}, \tag{5.35}$$

where h is the water depth, λ is the wavelength of the waves, and g is the acceleration due to gravity. This formula suggests that waves of different wavelengths will travel at different speeds. Waves with longer wavelengths travel faster than waves with shorter wavelengths.

For the waves in deep waters when the wavelength λ is much smaller than h (or $\lambda \ll h$), we have $\frac{2\pi h}{\lambda} \to \infty$, and $\exp[-\frac{2\pi h}{\lambda}] \to 0$. Now we get

$$\tanh(\frac{2\pi h}{\lambda}) = \frac{e^{\frac{2\pi h}{\lambda}} - e^{-\frac{2\pi h}{\lambda}}}{e^{\frac{2\pi h}{\lambda}} + e^{-\frac{2\pi h}{\lambda}}} \approx \frac{e^{\frac{2\pi h}{\lambda}} - 0}{e^{\frac{2\pi h}{\lambda}} + 0} \to 1, \tag{5.36}$$

which leads to

$$v = \sqrt{\frac{g\lambda}{2\pi} \tanh(\frac{2\pi h}{\lambda})} \approx \sqrt{\frac{g\lambda}{2\pi}}. \tag{5.37}$$

A tsunami is a giant water wave whose wavelength and speed are constantly changing as it travels towards the shore. In deep ocean waters, the wave height is typically less than half a metre, but its wavelength is in the range of 25 km to 50 km. Let us now estimate its speed in deep water using the typical values of $\lambda = 25$ km to 50 km (or 2.5×10^4 to 5.0×10^4 meters), $g = 9.8$ m/s^2. We have the phase speed

$$v = \sqrt{\frac{g\lambda}{2\pi}} = \sqrt{\frac{9.8 \times 2.5 \times 10^4}{2\pi}} \approx 197 \text{ m/s} \approx 444 \text{ mph}, \tag{5.38}$$

for $\lambda = 25$ km. For longer wavelength $\lambda = 50$ km, its speed is about 630 mph. This means that the first arrival of tsunami waves is always of long wavelength. As they travel shorewards, their wavelengths become typically in the range of 1.5 km to 5 km, but their wave heights can reach up to 30 metres. Their speed can reduce to 230 down to 25 mph. The total wave energy density E is measured by its amplitude A by $E = \frac{1}{8}\rho g A^2$ where ρ is the density of water. Therefore, a tsunami with $A = 30$ m will have tremendous power and could cause a lot of damage onshore.

In shallow waters where the wavelength λ is much longer than the depth h (that is $\lambda \gg h$), we have $\frac{2\pi h}{\lambda} \to 0$. Using $\exp(x) \approx 1+x$ as $x \to 0$ [see (6.34)] and $\tanh(x) = (e^x - e^{-x})/(e^x + e^{-x}) = (e^{2x} - 1)/(e^{2x} + 1)$, we have

$$\tanh(\frac{2\pi h}{\lambda}) = \frac{e^{2 \times \frac{2\pi h}{\lambda}} - 1}{e^{2 \times \frac{2\pi h}{\lambda}} + 1} \approx \frac{(1 + \frac{4\pi h}{\lambda}) - 1}{(1 + \frac{4\pi h}{\lambda}) + 1} \approx \frac{\frac{4\pi h}{\lambda}}{2 + \frac{4\pi h}{\lambda}} \approx \frac{2\pi h}{\lambda}. \tag{5.39}$$

Therefore, the speed of shallow water waves now becomes

$$v = \sqrt{\frac{g\lambda}{2\pi} \cdot \frac{2\pi h}{\lambda}} \approx \sqrt{gh}. \tag{5.40}$$

This simple formula can be used to estimate the water depth in a pond by dropping a pebble and recording how fast the induced ripples travel. For example, the ripples in a pond with a depth of $h = 0.5$ metres will travel at a speed of $v \approx \sqrt{9.8 \times 0.5} \approx 2.2$ m/s.

Chapter 6

Differentiation

Calculus is important in almost all quantitative sciences. In earth sciences, the calculations of deflection of plates under geological loading, the heat transfer of magma, and the understanding of the large-scale structures all use differentiation and integration. In this and the following chapters, we will introduce the fundamentals of differentiation and integration. We will also demonstrate their application through a few examples such as the path of a projectile, free-air gravity, and mantle convection under a constant temperature gradient.

6.1 Gradient

The gradient of a curve at any point P is the rate of change at that point, and it is the gradient of the tangent to the curve at that point. We know geometrically what a tangent means, but it is not easy to draw it accurately without calculations. In order to find the true gradient and the tangent, we normally use some adjustment points and try to use the ratio of the small change in y, $\delta y = y_Q - y_P$ to the change in x, $\delta x = x_Q - x_P$ to approximate the gradient. That is to say, we use

$$\frac{\delta y}{\delta x} = \frac{y_Q - y_P}{x_Q - x_P}, \tag{6.1}$$

to estimate the gradient. As Q is closer than R to P, thus the gradient estimated using P and Q is better than that using P and R (see Fig. 6.1). We hope that the estimate will become the true gradient as the point Q becomes very, very close to P. How do you describe such closeness? A simple way is to use the distance $h = \delta x = PS = x_Q - x_P$ and let h tend to zero. That is to say, $h \to 0$.

Since we know that the point $x_P = a$, its coordinates for the curve x^3 are (a, a^3). The adjustment point Q now has coordinates $(a+h, (a+$

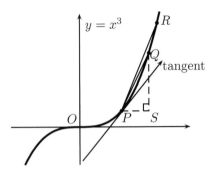

Figure 6.1: The tangent and gradient of a curve with $h=\mathrm{PS}=x_Q - x_P$.

$h)^3$). Therefore, the gradient at point P becomes

$$\frac{\delta y}{\delta x} = \frac{(a+h)^3 - a^3}{h}. \tag{6.2}$$

Since $(a+h)^3 = (a^3 + 3a^2h + 3ah^2 + h^3)$, we now have

$$\frac{\delta y}{\delta x} = \frac{(a^3 + 3a^2h + 3ah^2 + h^3) - a^3}{h} = 3a^2 + 3ah + h^2. \tag{6.3}$$

Since $h \to 0$, and also $h^2 \to 0$, both terms $3ah$ and h^2 will tend to zero. This means that the true gradient of x^3 at $x = a$ is $3a^2$. Since $x = a$ is any point, so in general the gradient of x^3 is $3x^2$.

Now let us introduce a more formal notation for the gradient of any curve $y = f(x)$. We define the gradient as

$$f'(x) \equiv \frac{dy}{dx} \equiv \frac{df(x)}{dx} = \lim_{h \to 0} \frac{f(x+h) - f(x)}{h}. \tag{6.4}$$

The gradient is also called the first derivative. The three notations $f'(x)$, dy/dx and $df(x)/dx$ are interchangeable.

Conventionally, the notation dy/dx is called Leibnitz's notation, while the prime notation $'$ is called Lagrange's notation. Newton's dot notation $\dot{y} = dy/dt$ is now exclusively used for time derivatives. The choice of such notations is purely for clarity, convention and/or personal preference.

In addition, here the notation 'lim' is used, as it is associated with the fact that $h \to 0$. The limit can be understood that for any $0 < h \le 1$ where h is much less than 1 and infinitesimally small, then there always exists another even smaller number $0 < \epsilon < h$ so that we can find a better estimate for the gradient using ϵ.

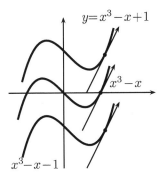

Figure 6.2: The gradients of a family of curves $y = x^3 - x + k$
(where $k = 0, \pm 1$) at $x = 1$ are the same $3x^2 - 1$.

From the basic definition of the derivative, we can verify that such
differentiation manipulation is a linear operator. That is to say that for
any two functions $f(x)$, $g(x)$ and two constants α and β, the derivative
or gradient of a linear combination of the two functions can be obtained
by differentiating the combination term by term. We have

$$[\alpha f(x) + \beta g(x)]' = \alpha f'(x) + \beta g'(x), \tag{6.5}$$

which can easily be extended to multiple terms. Furthermore, if k is a
real constant, then its derivative is zero. For example, if we have

$$\frac{d(x^3 - x + k)}{dx} = \frac{d(x^3)}{dx} - \frac{dx}{dx} + \frac{dk}{dx} = 3x^2 - 1 + 0 = 3x^2 - 1, \tag{6.6}$$

which means that for a family of curves shifted by a constant k, the
gradients at any same point (say $x = 1$) are all the same, and the
tangent lines are parallel to each other (see Fig. 6.2).

Example 6.1: To find the gradient $f(x) = x^n$ where $n = 1, 2, 3, 4, ...$
is a positive integer, we use

$$f'(x) = \lim_{h \to 0} \frac{f(x+h) - f(x)}{h} = \lim_{h \to 0} \frac{(x+h)^n - x^n}{h}.$$

Using the binomial theorem

$$(x+h)^n = \binom{n}{0} x^n + \binom{n}{1} x^{n-1} h + \binom{n}{2} x^{n-2} h^2 + ... + \binom{n}{n} h^n,$$

and the fact that

$$\binom{n}{0} = 1, \quad \binom{n}{1} = n, \quad \binom{n}{2} = \frac{n(n-1)}{2!}, \quad \binom{n}{k} = \frac{n!}{k!(n-k)!}, \quad \binom{n}{n} = 1,$$

we have

$$(x+h)^n - x^n = (x^n + nx^{n-1}h + \frac{n(n-1)}{2!}x^{n-2}h^2 + ... + h^n) - x^n$$

$$= nx^{n-1}h + \frac{n(n-1)}{2!}x^{n-2}h^2 + ... + h^n.$$

Therefore, we get

$$f'(x) = \lim_{h \to 0} \frac{nx^{n-1}h + \frac{n(n-1)}{2!}x^{n-2}h^2 + ... + h^n}{h}$$

$$= \lim_{h \to 0} \left[nx^{n-1} + \frac{n(n-1)}{2!}x^{n-2}h + ... + h^{n-1} \right].$$

Since $h \to 0$, all the terms except the first (nx^{n-1}) will tend to zero. We finally have

$$f'(x) = nx^{n-1}.$$

In fact, this formula is valid for any integer $n = 0, \pm 1, \pm 2, ...$

It is worth pointing out that the limit $\lim_{h \to 0}$ must exist. If such a limit does not exist, then the derivative or the gradient is not defined. For example, the gradient of a rather seemingly simple function $f(x) = |x|$ where $|x|$ means the absolute value of $f'(x)$ does not exist at $x = 0$ (see Fig. 6.3). This is because we will get different values depending on how we approach $x = 0$. If we approach $h \to 0$ from the right $(h > 0)$ and use the notation $h \to 0^+$, we have

$$f'(0^+) = \lim_{h \to 0^+} \frac{|h| - |0|}{h} = \lim_{h \to 0^+} \frac{h}{h} = \lim_{h \to 0^+} 1 = 1. \qquad (6.7)$$

However, if we approach from the left $(h < 0)$, we have

$$f'(0^-) = \lim_{h \to 0^-} \frac{|h| - |0|}{h} = \lim_{h \to 0^-} \frac{-h}{h} = \lim_{h \to 0^-} -1 = -1. \qquad (6.8)$$

The correct definition of the derivative should be independent of the way we approach the point using $h \to 0$. In this case, we say, the gradient or derivative at $x = 0$ does not exist.

We can obtain the derivative of any function from first principles. For example, we now try to derive $d(\sin x)/dx$. But before we can do this, we have to derive a formula for a small angle θ when $\theta \to 0$.

For simplicity, we now refer to Fig. 6.4. In the case when OA=1, we have AB=$\tan \theta$. Since OA=OP= 1, the area of the triangle OAP is half of the product of the base OA=1 and the height $h = 1 \times \sin \theta$, which is simply $\frac{1}{2} \sin \theta$. The area of a circle is πr^2, and the area of a sector of θ degrees is $\pi r^2 \times \frac{\theta}{360°} = \frac{\pi}{180°} \times \frac{1}{2}r^2\theta$ where θ is in degrees. If we

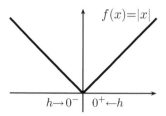

Figure 6.3: Limit and thus the derivative at $x = 0$ is not defined.

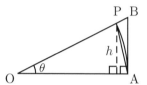

Figure 6.4: Small angle $\theta \to 0$ for OA=1.

simply use θ in radians with OA=1, the area of a sector OAP becomes $\frac{1}{2}\theta$. The area of triangle OAB is $\frac{1}{2} \times 1 \times \tan\theta$. From the geometrical point of view, we also observe that the area of triangle OAB is larger than the area of the sector OAP (of a circle with a radius $r = 1$=OA), which in turn is larger than the area of the triangle OAP. Therefore, we have

$$\frac{1}{2}\tan\theta > \frac{1}{2}\theta > \frac{1}{2}\sin\theta. \tag{6.9}$$

By multiplying by 2 and using $\tan\theta = \sin\theta/\cos\theta$, we get

$$\frac{\sin\theta}{\cos\theta} > \theta > \sin\theta. \tag{6.10}$$

Though θ is small but not zero, we can divide each term by $\sin\theta$, we obtain

$$\frac{1}{\cos\theta} > \frac{\theta}{\sin\theta} > 1. \tag{6.11}$$

As $\theta \to 0$, $\cos\theta \to 1$, we have $1 > \frac{\theta}{\sin\theta} > 1$, as $\theta \to 0$. That is to say,

$$\lim_{\theta \to 0}\frac{\theta}{\sin\theta} = 1, \qquad \text{or } \lim_{\theta \to 0}\frac{\sin\theta}{\theta} = 1. \tag{6.12}$$

Example 6.2:   In order to obtain the derivative of $f(x) = \sin(x)$, we now use the definition

$$f'(x) = \lim_{h \to 0}\frac{\sin(x+h) - \sin x}{h}.$$

From the formula (4.30), we know that

$$\sin(x+h) - \sin x = 2\cos\frac{[(x+h)+x]}{2}\sin\frac{[(x+h)-x]}{2}$$

$$= 2\cos(x+\frac{h}{2})\sin\frac{h}{2}.$$

We have

$$f'(x) = \lim_{h\to 0}\frac{2\cos(x+\frac{h}{2})\sin\frac{h}{2}}{h} = \lim_{h\to 0}\cos(x+\frac{h}{2})\frac{\sin\frac{h}{2}}{\frac{h}{2}}.$$

Using

$$\lim_{h\to 0}\frac{\sin\frac{h}{2}}{\frac{h}{2}} = 1, \qquad \lim_{h\to 0}\cos(x+\frac{h}{2}) = \cos x,$$

we finally have

$$f'(x) = \frac{d(\sin x)}{dx} = \cos x.$$

Following the same procedure, we can also derive that

$$\frac{d(\cos x)}{dx} = -\sin x,$$

which is left to the reader.

- -

This example demonstrates that even if we are able to calculate the derivative using first principles, it is usually not the best way to do so in practice. The good thing is that we only have to do it once to understand how it works. For more complicated functions, we should use certain rules, such as the chain rule. Even better, we can sometimes use tables or mathematical software packages.

6.2   Differentiation Rules

We know how to differentiate x^3 and $\sin(x)$, and a natural question is how we can differentiate $\sin(x^3)$? Here we need the chain rule. The trick to derive the chain rule is to consider dx and dy (and other similar quantities) as infinitesimal but non-zero quantities or increments δx and δy so that we can divide and multiply them as necessary. In the limit of $\delta x \to 0$, we have $\delta x \to dx$ and $\delta y/\delta x \to dy/dx$. If $f(u)$ is a function of u, and u is in turn a function of x, we want to calculate dy/dx. We then have

$$\frac{dy}{dx} = \frac{dy}{du}\cdot\frac{du}{dx}, \tag{6.13}$$

or

$$\frac{df[u(x)]}{dx} = \frac{df(u)}{du}\cdot\frac{du(x)}{dx}. \tag{6.14}$$

This is the well-known chain rule.

Example 6.3:   Now we come back to our original problem of differentiating $\sin(x^3)$. First we let $u = x^3$, we then

$$\frac{d(\sin(x^3))}{dx} = \frac{d(\sin u)}{du} \cdot \frac{dx^3}{dx} = \cos u \cdot 3x^2 = 3x^2 \cos(x^3).$$

As we practise more, we can write the derivatives more compactly and quickly. For example,

$$\frac{d(\sin^n(x))}{dx} = n \sin^{n-1}(x) \cdot \cos x,$$

where we consider $\sin^n(x)$ to be in the form of u^n and $u = \sin(x)$.

A further question is if we can differentiate $x^5 \sin(x)$ easily, since we already know how to differentiate x^5 and $\sin x$ separately? For this, we need the differentiation rule for products. Let $y = uv$ where $u(x)$ and $v(x)$ are (often simpler) functions of x. From the definition, we have

$$\frac{dy}{dx} = \frac{d(uv)}{dx} = \lim_{h \to 0} \frac{u(x+h)v(x+h) - u(x)v(x)}{h}. \tag{6.15}$$

Since adding a term and deducting the same term does not change an expression, we have

$$\frac{u(x+h)v(x+h) - u(x)v(x)}{h}$$

$$= \frac{u(x+h)v(x+h) - \overbrace{u(x+h)v(x) + u(x+h)v(x)}^{=0} - u(x)v(x)}{h}$$

$$= u(x+h)\frac{[v(x+h) - v(x)]}{h} + v(x)\frac{[u(x+h) - u(x)]}{h}. \tag{6.16}$$

In addition, $u(x+h) \to u(x)$ as $h \to 0$, we finally have

$$\frac{d(uv)}{dx} = \lim_{h \to 0} u(x+h)\frac{[v(x+h) - v(x)]}{h} + v(x)\frac{[u(x+h) - u(x)]}{h}$$

$$= u\frac{dv}{dx} + v\frac{du}{dx}, \tag{6.17}$$

or simply

$$(uv)' = uv' + vu'. \tag{6.18}$$

This is the formula for products or product rule. If we replace v by $1/v = v^{-1}$ and apply the chain rule

$$\frac{d(v^{-1})}{dx} = -1 \times v^{-1-1} \times \frac{dv}{dx} = -\frac{1}{v^2}\frac{dv}{dx}, \tag{6.19}$$

we have the formula for quotients or the quotient rule

$$\frac{d(\frac{u}{v})}{dx} = \frac{d(uv^{-1})}{dx} = u\left(\frac{-1}{v^2}\right)\frac{dv}{dx} + v^{-1}\frac{du}{dx} = \frac{v\frac{du}{dx} - u\frac{dv}{dx}}{v^2}. \qquad (6.20)$$

Now let us apply these rules to differentiating $\tan x$.

Example 6.4:   Since $\tan x = \sin x / \cos x$, we have $u = \sin x$ and $v = \cos x$. Therefore, we get

$$\frac{d(\tan x)}{dx} = \frac{\cos x \frac{d\sin x}{dx} - \sin x \frac{d\cos x}{dx}}{\cos^2 x} = \frac{\cos x \cos x - \sin x \times (-\sin x)}{\cos^2 x}$$

$$= \frac{\cos^2 x + \sin^2 x}{\cos^2 x} = \frac{1}{\cos^2 x}.$$

- -

Sometimes, it is not easy to find the derivative dy/dx directly. In this case, it is usually a good idea to try to find dx/dy since

$$\frac{dy}{dx} = 1 / \frac{dx}{dy}, \qquad (6.21)$$

or carry out the derivatives term-by-term and then find dy/dx. This is especially the case for implicit functions. Let us demonstrate this with an example.

Example 6.5:   To find the gradient of the curve

$$\sin^2 x + y^4 - 2y = e^{3x},$$

at the point $(0, 1)$, we first differentiate each term with respect to x, and we have

$$\frac{d\sin^2 x}{dx} + \frac{dy^4}{dx} - \frac{d(2y)}{dx} = \frac{de^{3x}}{dx},$$

or

$$2\sin x \cos(x) + 4y^3\frac{dy}{dx} - 2\frac{dy}{dx} = 3e^{3x},$$

where we have used the product rule

$$\frac{du}{dx} = \frac{du}{dy} \cdot \frac{dy}{dx},$$

so that $u = y^4$ gives

$$\frac{dy^4}{dx} = 4y^3\frac{dy}{dx}.$$

After some rearrangement, we have

$$(4y^3 - 2)\frac{dy}{dx} = 3e^{3x} - 2\sin x \cos x,$$

or

$$\frac{dy}{dx} = \frac{3e^{3x} - 2\sin x \cos x}{(4y^3 - 2)}.$$

Therefore, the gradient at $x = 0$ and $y = 1$ becomes

$$\frac{dy}{dx} = \frac{3e^{3\times 0} - 2 \times \sin 0 \times \cos 0}{4 \times 1^3 - 2} = \frac{3}{2} = 1.5.$$

The first derivatives of commonly used mathematical functions are listed in Table 6.1, and the derivatives of other functions can be obtained using the differentiation rules in combination with this table.

Table 6.1: First Derivatives of common functions.

$f(x)$	$f'(x)$	$f(x)$	$f'(x)$
x^n	nx^{n-1}	e^x	e^x
a^x	$a^x \ln a \ (a>0)$	$\ln x$	$\frac{1}{x} \ (x > 0)$
$\sin x$	$\cos x$	$\cos x$	$-\sin x$
$\tan x$	$1 + \tan^2 x$	$\sec x$	$\sec x \tan x$
$\log_a x$	$\frac{1}{x \ln a}$	$\tanh x$	$\text{sech}^2 x$
$\sinh x$	$\cosh x$	$\cosh x$	$\sinh x$
$\sin^{-1} x$	$\frac{1}{\sqrt{1-x^2}}$	$\cos^{-1} x$	$\frac{-1}{\sqrt{1-x^2}}$
$\tan^{-1} x$	$\frac{1}{1+x^2}$	$\sinh^{-1} x$	$\frac{1}{\sqrt{x^2+1}}$
$\tanh^{-1} x$	$\frac{1}{1-x^2}$	$\cosh^{-1} x$	$\frac{1}{\sqrt{x^2-1}}$

The derivative we discussed so far is the gradient or the first derivative of a function $f(x)$. If we want to see how the gradient itself varies, we may need to take the gradient of the gradient of a function. In this case, we are in fact calculating the second derivative. We write

$$\frac{d^2 f(x)}{dx^2} \equiv f''(x) \equiv \frac{d}{dx}\left(\frac{df(x)}{dx}\right). \tag{6.22}$$

Following a similar procedure, we can write any higher-order derivative as

$$\frac{d^3 f(x)}{dx^3} = f'''(x) = \frac{d(f''(x))}{dx}, \dots \tag{6.23}$$

For example, $f(x) = x^3$, we have $f'(x) = 3x^2$, $f''(x) = d(3x^2)/dx = 3 \times 2x = 6x$, $f'''(x) = 6$ and $f''''(x) = 0$. It is worth pointing out that notation for the second derivative is $\frac{d^2 y}{dx^2}$, not $\frac{d^2 y}{d^2 x}$ or $\frac{dy^2}{dx^2}$ which is wrong.

6.3 Partial Derivatives

For a smooth curve, it is relatively straightforward to draw a tangent line at any point; however, for a smooth surface, we have to use the tangent plane (see Fig. 6.5). For example, we now want to take the derivative of a function of two independent variables x and y, that is $z = f(x,y) = x^2 + y^2/2$. The question is probably 'with respect to' what? x or y? If we take the derivative to x, then will it be affected by y? The answer is we can take the derivative with respect to either x or y while taking the other variable as constant. That is, we can calculate the derivative with respect to x in the usual sense by assuming that $y = $ constant. Since there is more than one variable, we have more than one derivative and the derivatives can be associated with either the x-axis or y-axis. We call such derivatives partial derivatives, and use the following notation

$$\frac{\partial z}{\partial x} \equiv \frac{\partial f(x,y)}{\partial x} \equiv f_x \equiv \frac{\partial f}{\partial x}\bigg|_y = \lim_{h \to 0, y=\text{const}} \frac{f(x+h,y) - f(x,y)}{h}.$$
(6.24)

The notation $\frac{\partial f}{\partial x}\big|_y$ emphasises the fact that $y = $ constant; however, we often omit $\big|_y$ and simply write $\frac{\partial f}{\partial x}$ as we know this fact is implied. Similarly, the partial derivative with respect to y is defined by

$$\frac{\partial z}{\partial y} \equiv \frac{\partial f(x,y)}{\partial y} \equiv f_y \equiv \frac{\partial f}{\partial y}\bigg|_x = \lim_{x=\text{const}, k \to 0} \frac{f(x, y+k) - f(x,y)}{k}. \quad (6.25)$$

Then, the standard differentiation rules for univariate functions such as $f(x)$ apply. For example, for $z = f(x,y) = x^2 + y^2/2$, we have

$$\frac{\partial f}{\partial x} = \frac{dx^2}{dx} + 0 = 2x,$$

and $\frac{\partial f}{\partial y} = 0 + \frac{d(y^2/2)}{dy} = \frac{1}{2} \times 2y = y$, where 0 highlight the fact that $dy/dx = dx/dy = 0$ as x and y are independent variables.

Other than considering the other variable to be a constant, all the differentiation rules are the same. Let us look at a more complicated example by using the chain and product rules.

Example 6.6:   For $z = f(x,y) = \sin(xy^2) + x^2 y \sin x + y^5 \cos x$, we first calculate f_x while thinking $y = $ constant, and we have

$$f_x = \frac{\partial f}{\partial x} = \cos(xy^2) \times y^2 + y(2x \sin x + x^2 \cos x) + y^5(-\sin x)$$

$$= y^2 \cos(xy^2) + 2xy \sin x + x^2 y \cos x - y^5 \sin x.$$

Similarly, we have $\frac{\partial f}{\partial y} = \cos(xy^2) \times 2yx + x^2 \sin x + 5y^4 \cos x$, which is $f_y = 2xy \cos(xy^2) + x^2 \sin x + 5y^4 \cos x$.

We know that for a univariate function $f(x)$, we can calculate the change in Δf for any given small increment Δx. That is $\Delta f \approx \frac{df}{dx} \Delta x =$

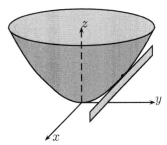

Figure 6.5: Partial derivatives of $f(x, y) = x^2 + y^2/2$.

$f'\Delta x$. Now for a function $z(x, y)$ of two independent variables x and y, we have to account for the increments in both x and y. In this case, we have to use the total differential which can be defined by

$$\Delta z \approx \frac{\partial z}{\partial x} \Delta x + \frac{\partial z}{\partial y} \Delta y, \tag{6.26}$$

where Δx and Δy are the increments of x and y. This formula is very useful for finding the propagation of errors.

6.4 Maxima or Minima

We know that the gradient of a curve at any point is the rate of change at that point. This means that the first derivative provides a vivid description of the variations along the curve. In Fig. 6.6, we can see that the curve has three points A, B and C with zero-gradient (three horizontal lines). We say that these points are stationary points if at these points the gradient or the first derivative of the curve is zero, that is $f'(x) = 0$, since the rate of change is zero and y here is essentially stationary.

Around these stationary points, the gradient may change signs. For example, at point A, the gradient changes from the positive on the left to negative on the right. This suggests that the gradient itself is changing substantially, and the derivative of the gradient is not zero. Therefore, any information about the second derivative $f''(x)$ may be useful. At point A, the gradient change from positive to negative suggests that the second derivative is negative or $f''(x) < 0$. Both conditions $f'(x) = 0$ and $f''(x) < 0$ at point A imply that A is a local maximum point. These same conditions apply at point C. On the other hand, at point B, the stationary condition $f'(x) = 0$ is the same, but the gradient changes from the negative on the left to positive on the right, which suggests that the second derivative $f''(x) > 0$, corresponding to

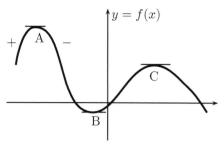

Figure 6.6: Stationary points, minimum and maximum.

a local minimum. Any minimum or maximum point is called a turning point, so all points A, B, C are also turning points.

However, $f'(x) = 0$ does not necessarily always correspond to a turning point. For example, the point O of the curve $y = x^3$ in Fig. 6.1 is a stationary point as $f'(x) = 3x^2 = 0$ at $x = 0$, but $f''(x) = 6x = 0$ at $x = 0$, so it is not a turning point. It is in fact an inflexion point. An inflexion point is a point on a curve where $f''(x) = 0$, and the gradient at a point of inflexion is not necessarily zero.

In many applications, sometimes the main aim is to find the minima or maxima of a curve as the solution.

Example 6.7:   For example, if we want to throw a small object at an initial velocity v with an angle θ above the horizontal direction, what is the farthest possible distance d and what is the height H (see Fig. 6.7)?

From basic physics, the equation of the projectile can be written as

$$y = x \tan \theta - \frac{g}{2v^2 \cos^2 \theta} x^2.$$

At the highest point A, y becomes stationary, that is $dy/dx = 0$ or $\frac{dy}{dx} = \tan \theta - \frac{gx}{v^2 \cos^2 \theta} = 0$. This gives

$$x_* = \frac{v^2 \cos^2 \theta \, \tan \theta}{g} = \frac{v^2 \cos^2 \theta \frac{\sin \theta}{\cos \theta}}{g} = \frac{v^2 \sin \theta \cos \theta}{g} = \frac{v^2 \sin 2\theta}{2g},$$

where we have used $\sin 2\theta = 2 \sin \theta \cos \theta$. The distance is $d = 2x_*$

$$d = 2x_* = \frac{v^2 \sin 2\theta}{g}.$$

The height H is

$$H = x_* \tan \theta - \frac{g}{2v^2 \cos^2 \theta} x_*^2 = (\frac{v^2 \sin 2\theta}{2g}) \tan \theta - \frac{g}{2v^2 \cos^2 \theta} (\frac{v^2 \sin 2\theta}{2g})^2$$

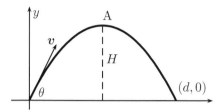

Figure 6.7: Projectile of an object with an initial velocity v at an angle θ.

$$= \left(\frac{2v^2 \sin \theta \cos \theta}{2g}\right)\frac{\sin \theta}{\cos \theta} - \frac{v^4 \sin^2 \theta \cos^2 \theta}{2v^2 g \cos^2 \theta} = \frac{v^2 \sin^2 \theta}{2g}.$$

In order to reach the farthest distance d, we can also adjust the angle θ for a given velocity v so that $\sin 2\theta = 1$ reaches its maximum. That is $2\theta = 90°$ or $\theta = 45°$.

On the other hand, if we do not care about the distance, but want to reach the highest possible height H, then we can adjust the angle so that $\sin^2 \theta = 1$ or $\theta = 90°$. That is to throw vertically.

- -

6.5 Series Expansions

In numerical methods and some mathematical analysis, series expansions make some calculations easier. For example, we can write the exponential function e^x as a series about $x_0 = 0$

$$e^x = \alpha_0 + \alpha_1 x + \alpha_2 x^2 + \alpha_3 x^3 + ... + \alpha_n x^n. \tag{6.27}$$

Now let us try to determine these coefficients. At $x = 0$, we have

$$e^0 = 1 = \alpha_0 + \alpha_1 \times 0 + \alpha_2 \times 0^2 + ...\alpha_n \times 0^n = \alpha_0, \tag{6.28}$$

which gives $\alpha_0 = 1$. In order to reduce the power or order of the expansion so that we can determine the next coefficient, we first differentiate both sides of (6.27) once; we have

$$e^x = \alpha_1 + 2\alpha_2 x + 3\alpha_3 x^2 + ... + n\alpha_n x^{n-1}. \tag{6.29}$$

By setting again $x = 0$, we have

$$e^0 = 1 = \alpha_1 + 2\alpha_2 \times 0 + ... + n\alpha_n \times 0^{n-1} = \alpha_1, \tag{6.30}$$

which gives $\alpha_1 = 1$. Similarly, differentiating it again, we have

$$e^x = (2 \times 1) \times \alpha_2 + 3 \times 2\alpha_3 x + ... + n(n-1)\alpha_n x^{n-2}. \tag{6.31}$$

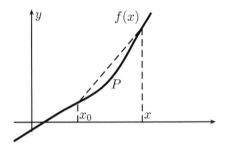

Figure 6.8: Expansion and approximations for $f(x) = f(x_0+h)$ where $h = x - x_0$.

At $x = 0$, we get

$$e^0 = (2 \times 1) \times \alpha_2 + 3 \times 2\alpha_3 \times 0 + ... + n(n-1)\alpha_n \times 0^{n-2} = 2\alpha_2, \quad (6.32)$$

or $\alpha_2 = 1/(2 \times 1) = 1/2!$. Following the same procedure and differentiating it n times, we have

$$e^x = n!\alpha_n, \quad (6.33)$$

and $x = 0$ leads to $\alpha_n = 1/n!$. Therefore, the final series expansion can be written as

$$e^x = 1 + x + \frac{1}{2!}x^2 + \frac{1}{3!}x^3 + ... + \frac{1}{n!}x^n. \quad (6.34)$$

Obviously, we can follow a similar process to expand other functions. We have seen here the importance of differentiation and derivatives.

If we know the value of $f(x)$ at x_0, we know that the gradient $f'(x)$ in a small interval $h = x - x_0$ can be approximated by (see Fig. 6.8)

$$f'(x) = \frac{f(x_0 + h) - f(x_0)}{h}, \quad (6.35)$$

which is equivalent to saying that the adjacent value of $f(x_0 + h)$ can be approximated by

$$f(x_0 + h) \approx f(x_0) + hf'(x). \quad (6.36)$$

Then as a first approximation we can use $f'(x_0)$ to approximate $f'(x)$ in the interval. However, if we know higher derivatives $f''(x_0)$ at x_0, we can use an even better approximation to estimate $f'(x)$

$$f'(x_0 + h) = f'(x_0) + h\frac{f''(x)}{2}. \quad (6.37)$$

This is equivalent to stating that

$$f(x_0 + h) = f(x_0) + h f'(x_0) + \frac{h^2}{2} f''(x). \qquad (6.38)$$

If we again approximate $f''(x)$ using higher-order derivatives, we can obtain the Taylor series expansions

$$f(x_0 + h) = f(x_0) + f'(x_0)h + \frac{f''(x_0)}{2!} h^2$$

$$+ \frac{f'''(x_0)}{3!} h^3 + \dots + \frac{f^{(n)}(x_0)}{n!} h^n + R_{n+1}(h), \qquad (6.39)$$

where $f^{(n)}$ is the nth derivative of $f(x)$ and $R_{n+1}(h)$ is the error estimation for this series. In theory, we can use as many terms as possible, but in practice, the series converges very quickly and only a few terms are sufficient. It is straightforward to verify that the exponential series for e^x is identical to the results given earlier. Now let us look at other examples.

Example 6.8:   Let us expand $f(x) = \sin x$ about $x_0 = 0$. We know that

$$f'(x) = \cos x, \quad f''(x) = -\sin x, \quad f'''(x) = -\cos x, \quad \dots,$$

or $f'(0) = 1$, $f''(0) = 0$, $f'''(0) = -1$, $f''''(0) = 0, \dots$, which means that

$$\sin x = \sin 0 + x f'(0) + \frac{f''(0)}{2!} x^2 + \frac{f'''(0)}{3!} x^3 + \dots = x - \frac{x^3}{3!} + \frac{x^5}{5!} + \dots,$$

where the angle x is in radians.

For example, we know that $\sin 30° = \sin \frac{\pi}{6} = 1/2$. We now use the expansion to estimate it for $x = \pi/3 = 0.523598$

$$\sin \frac{\pi}{6} \approx \frac{\pi}{6} - \frac{(\pi/6)^3}{3!} + \frac{(\pi/6)^5}{5!}$$

$$= 0.523599 - 0.02392 + 0.0000328 \approx 0.5000021326,$$

which is very close to the true value $1/2$.

6.6   Applications

6.6.1   Free-Air Gravity

Let us use direct differentiation to derive the gradient of gravity for free-air gravity correction.

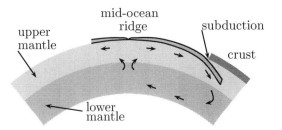

Figure 6.9: Mantle convection and subduction zone.

We know the gravity g on the Earth's surface is

$$g = \frac{GM_E}{r^2}, \tag{6.40}$$

where M_E is the mass of the Earth, and r is the radius of the Earth. G is the universal gravitational constant. The gradient of gravity is the first derivative with respect to r; we have

$$\frac{dg}{dr} = GM_E \frac{d}{dr}(r^{-2}) = GM_E(-2)r^{-2-1} = -\frac{2GM_E}{r^3} = -\frac{2g}{r}. \tag{6.41}$$

Since $g = 9.8$ m/s^2 and $r = 6370$ km $= 6.37 \times 10^6$ m, we have

$$\frac{dg}{dr} = -\frac{2 \times 9.8}{6.37 \times 10^6} \approx -3.08 \times 10^{-6} \text{ m/s}^2 \text{ m} = -0.308 \text{ mGal/m},$$

where we have used geophysical units: 1 mGal $= 10^{-3}$ Gal $= 10^{-5}$ m/s^2 and 1 Gal $= 1$ cm/s^2.

Therefore, if we go up to a height h, the free-air correction, or the difference between the gravity at h and that on the ground, should be

$$\Delta g_h = -0.308h \text{ mGal/m}, \tag{6.42}$$

which is well-known in gravity survey where the average gradient -0.3086 mGal/m is often used.

6.6.2 Mantle Convection

The main driving mechanism for plate tectonics is mantle convection, arising from the temperature gradient above a certain critical level under the right conditions. In such largescale convection, hot upwelling fluid goes up at the mid-ocean ridge and cold downgoing current subducts at the subduction zone (see Fig. 6.9).

For vigorous convection, convecting fluid is virtually adiabatic and the entropy S of the system can be considered as constant. Thermodynamics analysis suggests that the adiabatic thermal gradient at constant entropy is given by

$$\frac{dT}{dp} = \frac{T\alpha}{\rho c_p}, \tag{6.43}$$

where T is the average temperature, p is the pressure and α is the thermal expansion coefficient. ρ and c_p are the density and specific heat of the convecting fluid, respectively. If we are concerned with vertical gradient only, as in the case of mantle convection, temperature varies with depth z. Using the chain rule of differentiation, we have

$$\frac{dT}{dp} = \frac{\partial T}{\partial z}\frac{\partial z}{\partial p} = \frac{\partial T}{\partial z} / \frac{\partial p}{\partial z}. \tag{6.44}$$

Assuming that the pressure is hydrostatic $p = \rho g z$ so that $dp/dz = \rho g$ where g is the acceleration due to gravity, and using (6.43), we have the adiabatic thermal gradient

$$\frac{dT}{dz} = \frac{dT}{dp}\frac{dp}{dz} = \frac{T\alpha}{\rho c_p}\frac{dp}{dz} = \frac{T\alpha g}{c_p}. \tag{6.45}$$

Using typical values for upper mantle $T = 1600$ K, $\alpha = 3 \times 10^{-5}$ K^{-1}, $c_p = 1000$ J K^{-1} kg^{-1} and $g = 9.8$ m/s^2, we have

$$\frac{dT}{dz} = \frac{1600 \times 3 \times 10^{-5} \times 9.8}{1000} \approx 0.47 \times 10^{-3} \text{ K/m} = 0.47 \text{ K/km}. \tag{6.46}$$

This means that the temperature change in the 3000 km mantle is just $0.47 \times 3000 \approx 1400$ K.

Fluid dynamics has demonstrated that a horizontal fluid layer heated from below remains quiescent until the thermal gradient exceeds a critical value, resulting in cellular convection, called Rayleigh-Bénard convection. For mantle convection, the dimensionless Rayleigh number Ra is defined by

$$Ra = \frac{\rho g \alpha \Delta T d^3}{\kappa \mu} = \frac{g \alpha \Delta T d^3}{\kappa \nu}, \tag{6.47}$$

where $\nu = \mu/\rho$ is the kinematic viscosity and μ is the absolute or dynamic viscosity. κ is the thermal diffusivity, ΔT is the temperature difference across the convection layer with a thickness of d, and ρ is the average density of the mantle. Convection occurs if Ra is greater than a critical value $Ra_* \approx 27\pi^4/4 \approx 657.5$ derived from instability analysis, that is $Ra > Ra_*$. Viscosity in mantle varies greatly from $10^{19} \sim 10^{20}$ Pa s in the upper mantle to $10^{20} \sim 10^{23}$ Pa s in the lower mantle. Using the typical values $\mu = 10^{21}$ Pa s, $\Delta T = 1400$ K, $\rho = 3700$

kg/m^3, $\kappa = 10^{-6} \text{ m}^2/\text{s}$, $d = 3000 \text{ km} = 3 \times 10^6$ m, and the values for α and g given earlier, we have

$$Ra = \frac{3700 \times 9.8 \times 3 \times 10^{-5} \times 1400 \times (3 \times 10^6)^3}{10^{-6} \times 10^{21}} \approx 4 \times 10^7. \quad (6.48)$$

Such a high Rayleigh number implies that mantle convection is rigorous. For Rayleigh-Bénard convection, the horizontal size of a convection cell is similar to its vertical thickness. However, many studies on mantle convection suggest that the horizontal size is about $10,000$ km while the vertical size varies from 670 km to 3000 km. This implies that the viscosity in the lower mantle should be much higher than that in the upper mantle. The detailed mechanism of mantle convection is still poorly understood, and it is still an active area of current research.

Using a simple boundary layer theory with appropriate approximations, we know that the convection flow is more rigorous with an averaged convection velocity v inside a boundary layer with a thickness of w. The total buoyancy force of the downgoing current is approximately

$$F_b = \rho g \alpha (\frac{\Delta T}{2}) w A, \quad (6.49)$$

where $\Delta T/2$ is the average temperature difference, and A is the horizontal area of the boundary layer. Since the shear stress of the layer is $\tau = \mu(v/w)$ where v/w is the strain rate, the total viscous drag force is

$$F_d = \tau A = \frac{\mu v}{w} A. \quad (6.50)$$

In the steady-state convection, the two forces should be balanced $F_b = F_d$, and we have

$$\rho g \alpha (\frac{\Delta T}{2}) w A = \frac{\mu v}{w} A, \quad (6.51)$$

or

$$v = \frac{\rho g \alpha \Delta T w^2}{2\mu}. \quad (6.52)$$

Detailed boundary-layer analysis (Turcotte and Schubert, 1982; Richards et al., 2000) suggests that $w = \sqrt{\kappa d/v}$. We now have the velocity of mantle convection

$$v = \sqrt{\frac{\rho g \alpha \Delta T \kappa d}{2\mu}}. \quad (6.53)$$

Using the typical values given earlier, we have

$$v = \sqrt{\frac{3700 \times 9.8 \times 3 \times 10^{-5} \times 1400 \times 10^{-6} \times (3 \times 10^6)}{2 \times 10^{21}}} \approx 1.5 \times 10^{-9} \text{ m/s},$$

which is about 5 cm/year since 1.5×10^{-9} (m/s) $\times 365 \times 24 \times 3600$ (s/year) ≈ 5 cm/year. This example demonstrates that the use of appropriate values for various parameters is crucially important to obtain a realistic estimate.

Chapter 7

Integration

Integration is another important topic in calculus. It can be considered as the reverse of differentiation. Loosely speaking, differentiation and integration are two sides of the same coin. Integrals are important in finding areas, solving differential equations, signal processing and many branches of earth sciences such as geodynamics. Later in this chapter, we will use integration to study unit hydrograph and derive the formula for Bouguer gravity corrections.

7.1   Integration

Differentiation is used to find the gradient for a given function. Now a natural question is how to find the original function for a given gradient. This is the integration process, which can be considered as the reverse of the differentiation process. Since we know that

$$\frac{d\sin x}{dx} = \cos x, \qquad (7.1)$$

that is the gradient of $\sin x$ is $\cos x$, we can easily say that the original function is $\sin x$ if we know its gradient is $\cos x$. We can write

$$\int \cos x \, dx = \sin x + C, \qquad (7.2)$$

where C is the constant of integration. Here $\int \, dx$ is the standard notation showing the integration is with respect to x, and we usually call this the integral. The function $\cos x$ is called the integrand.

The integration constant comes from the fact that a family of curves shifted by a constant will have the same gradient at their corresponding points (see Fig. 6.2). This means that the integration can be determined up to an arbitrary constant. For this reason, we call it the indefinite integral.

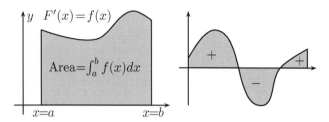

Figure 7.1: Integration and geometrical interpretation.

Integration is more complicated than differentiation in general. Even when we know the derivative of a function, we have to be careful. For example, we know that $(x^{n+1})' = (n+1)x^n$ or $(\frac{1}{n+1}x^{n+1})' = x^n$ for any n integers, so we can write

$$\int x^n dx = \frac{1}{n+1}x^{n+1} + C. \tag{7.3}$$

However, there is a possible problem when $n = -1$ because $1/(n+1)$ will become $1/0$. In fact, the above integral is valid for any n except $n = -1$. When $n = -1$, we have

$$\int \frac{1}{x}dx = \ln x + C. \tag{7.4}$$

If we know that the gradient of a function $F(x)$ is $f(x)$ or $F'(x) = f(x)$, it is possible and sometimes useful to express where the integration starts and ends, and we often write

$$\int_a^b f(x)dx = \Big[F(x)\Big]_a^b = F(b) - F(a). \tag{7.5}$$

Here a is called the lower limit of the integration, while b is the upper limit of the integration. In this case, the constant of integration has dropped out because the integral can be determined accurately. The integral becomes a definite integral and it corresponds to the area under a curve $f(x)$ from a to $b \geq a$ (see Fig. 7.1).

It is worth pointing out that the area in Fig. 7.1 may have different signs depending on whether the curve $f(x)$ is above or below the x-axis. If $f(x) > 0$, then that section of area is positive, otherwise, it is negative.

From (7.5), if we interchange the limits of the integral, we have

$$\int_b^a f(x)dx = \Big[F(x)\Big]_b^a = F(a) - F(b) = -(F(b) - F(a)) = -\int_a^b f(x)dx. \tag{7.6}$$

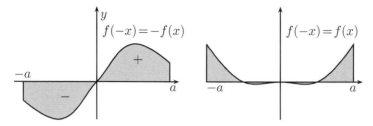

Figure 7.2: Integrals of odd and even functions.

In addition, since differentiation is linear, that is $[\alpha f(x) + \beta g(x)]' = \alpha f'(x) + \beta g'(x)$ where α and β are two real constants, it is easy to understand that integration is also a linear operator. We have

$$\int_a^b [\alpha f(x) + \beta g(x)]dx = \alpha \int_a^b f(x)dx + \beta \int_a^b g(x)dx. \qquad (7.7)$$

Also if $a \leq c \leq b$, we have

$$\int_a^b f(x)dx = \int_a^c f(x)dx + \int_c^b f(x)dx. \qquad (7.8)$$

This may become useful when dealing with certain integrals, especially if we do not want the areas (positive and negative) to simply add together.

In the special case of an odd function $f(-x) = -f(x)$, we have

$$\int_{-a}^a f(x)dx = \int_{-a}^0 f(x)dx + \int_0^a f(x)dx = -\int_0^{-a} f(x)dx + \int_0^a f(x)dx$$

$$= \int_0^a f(-x)dx + \int_0^a f(x)dx = \int_0^a [-f(x) + f(x)]dx = 0. \qquad (7.9)$$

The other way of understanding this is that the area for $x > 0$ is opposite in sign to the area for $x < 0$. So the total area or the integral is zero.

Similarly, for an even function $f(-x) = f(x)$, the areas for both $x > 0$ and $x < 0$ are the same, so we have

$$\int_{-a}^a f(x)dx = 2 \int_0^a f(x)dx. \qquad (7.10)$$

Example 7.1:   Let us try to compute the total area of the shaded regions under the curve $f(x) = x^2 - 5x + 4$ shown in Fig. 7.3.

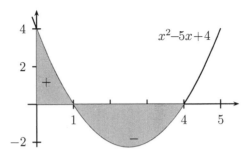

Figure 7.3: Computing the total area of the shaded regions.

First, we can simply say that the area is obviously the definite integral

$$A = \int_0^4 f(x)dx = \int_0^4 (x^2 - 5x + 4)dx.$$

From the basic linear properties of the integration, we have

$$A = \int_0^4 x^2 dx - 5\int_0^4 x dx + 4\int_0^4 dx = \left[\frac{1}{3}x^3\right]_0^4 - 5\left[\frac{1}{2}x^2\right]_0^4 + 4\left[x\right]_0^4$$

$$= \frac{1}{3} \times [4^3 - 0^3] - \frac{5}{2} \times [4^2 - 0^2] + 4 \times [4 - 0] = \frac{64}{3} - 40 + 16 = -\frac{8}{3}.$$

The total area is negative; this is not what we expected. The area should be positive, right? The geometrical meaning of the integral is the total area, adding the positive areas (above the x-axis) and the negative (below the x-axis). However, what we wanted to calculate is simply the total area of the two shaded regions, not to include its signs. In this case, we have to divide the integral into two intervals $[0, 1]$ and $[1, 4]$. The area of the first region is

$$\int_0^1 f(x)dx = \frac{1}{3} \times [1^3 - 0^3] - \frac{5}{2} \times [1^2 - 0^2] + 4 \times [1 - 0] = \frac{11}{6}.$$

Similarly, the area of the second region is

$$\int_1^4 f(x)dx = \frac{1}{3} \times [4^3 - 1^3] - \frac{5}{2} \times [4^2 - 1^2] + 4 \times [4 - 1]$$

$$= \frac{63}{3} - \frac{75}{2} + 12 = -\frac{9}{2}.$$

Here we only want the value of the area, so the area of the second region is $9/2$. Therefore, the total area of the shaded regions is $A = \frac{11}{6} + \frac{9}{2} = 6\frac{1}{3}$.

The integrals of commonly used functions are listed in Table 7.1. For many simple functions, we can easily write their integrals using the table of differentiation. For more complicated functions, what are the general techniques to obtain integrals? There are many techniques such as integration by parts, method of substitution, complex integrals and others. In the next section, we will briefly introduce integration by parts.

Table 7.1: Integrals of common functions.

$f(x)$	$\int f(x)dx$	$f(x)$	$\int f(x)dx$
e^x	e^x	$\frac{1}{x}$	$\ln x$
x^n	$\frac{1}{n+1}x^{n+1}$ $(n \neq -1)$	a^u	$\frac{a^u}{\ln a}$ $(a > 0)$
$\ln x$	$x \ln x - x$	$\log_a x$	$x \log_a x - \frac{x}{\ln a}$
$\frac{1}{a^2+x^2}$	$\frac{1}{a} \tan^{-1} \frac{x}{a}$	$\frac{1}{a^2-x^2}$	$\frac{1}{2a} \ln \frac{a+x}{a-x}$
$\frac{1}{x^2-a^2}$	$\frac{1}{2a} \ln \frac{x-a}{x+a}$	$\frac{1}{\sqrt{a^2-x^2}}$	$\sin^{-1} \frac{x}{a}$
$\frac{1}{\sqrt{x^2+a^2}}$	$\sinh^{-1} \frac{x}{a}$	$\frac{1}{\sqrt{x^2-a^2}}$	$\cosh^{-1} \frac{x}{a}$
$\sin x$	$-\cos x$	$\cos x$	$\sin x$
$\sec^2 x$	$\tan x$	$\csc^2 x$	$-\cot x$
$\sinh x$	$\cosh x$	$\cosh x$	$\sinh x$
$\tanh x$	$\ln \cosh x$	$\coth x$	$\ln \sinh x$

Example 7.2:   Horton's equation is an empirical relationship concerning the infiltration rates in groundwater flow. In the simplest case, the infiltration rate $f(t)$ at time t takes the form

$$f(t) = f_* + (f_0 - f_*)e^{-\beta t},$$

where f_0 is the maximum infiltration rate, and f_* is the infiltration rate at the equilibrium after the soil has been saturated. The parameter β is constant and $\tau = 1/\beta$ defines a characteristic timescale for infiltration.

In order to calculate the total volume of infiltration from time 0 to T, we take the integral

$$V = \int_0^T f(t)dt = \int_0^T [f_* + (f_0 - f_*)e^{-\beta t}]dt$$

$$= f_*T + (f_0 - f_*)\left[\frac{-1}{\beta}e^{-\beta t}\right]_0^T$$

$$= f_*T + \frac{(f_0 - f_*)}{\beta}(1 - e^{-\beta T}).$$

7.2   Integration by Parts

In differentiation, we can easily find the gradient of $x \sin x$ if we know the derivatives of x and $\sin x$ by using the rule of products. In integration, can we find the integral of $x \sin x$ if we know the integrals of x and $\sin x$? The answer is yes, this is the integration by parts.

From the differentiation rule

$$\frac{d(uv)}{dx} = u\frac{dv}{dx} + v\frac{du}{dx}, \tag{7.11}$$

we integrate it with respect to x, we have

$$uv = \int u\frac{dv}{dx}dx + \int v\frac{du}{dx}dx. \tag{7.12}$$

By rearranging, we have

$$\int u\frac{dv}{dx}dx = uv - \int v\frac{du}{dx}dx. \tag{7.13}$$

This is the well-known formula for the technique known as integration by parts. You may wonder where the constant of integration is? You are right, there is a constant of integration for uv, but as we know they exist for indefinite integrals, we simply omit to write it out in the formula, but we have to remember to put it back in the end. Now let us look at a simple example.

Example 7.3:   A unit hydrograph is an important tool in hydrology. There are many different methods and formulae to approximate the unit hydrographs such as the triangular representations. From the mathematical point of view, the following form can fit well

$$f(t) = At^n e^{-t/\tau},$$

where A is a scaling constant, and τ is the time constant. The exponent $n > 0$ and τ as well as A can be obtained by fitting to the experimental data.

The total discharge from 0 to $t = T$ can be calculated by the integral

$$Q = \int_0^T f(t)dt.$$

Let us consider a special case of $n = 1$ for simplicity. Now we have

$$Q = A\int_0^T te^{-t/\tau}dt.$$

Setting $u = t$ and $v' = e^{-t/\tau}$ and using integration by part, we have

$$Q = A\{\left[t(-\tau)e^{-t/\tau}\right]_0^T - \int_0^T e^{-t/\tau}(-\tau)dt\}$$

$$= A\{-\tau T e^{-T/\tau} + \tau\left[(-\tau)e^{-t/\tau}\right]_0^T\} = A\tau^2 - A\tau(\tau + T)e^{-T/\tau}.$$

We can see that

$$Q \to A\tau^2, \qquad \text{as} \quad T \to \infty.$$

So for a given discharge Q, the characteristic discharge time is

$$\tau = \sqrt{\frac{Q}{A}}.$$

When applying the technique of integration by parts, it might be helpful to pause and look at different perspectives, and the solutions sometimes come out more naturally. For example, we try to find the integral

$$I = \int e^x \sin x \, dx. \tag{7.14}$$

If we use $u = e^x$ and $dv/dx = \sin x$, we have $du/dx = e^x$ and $v = -\cos x$. By using integration by parts, we then have

$$I = \int e^x \sin x \, dx = e^x(-\cos x) - \int (-\cos x)\frac{du}{dx} dx$$

$$= -\cos x e^x + \int e^x \cos x \, dx. \tag{7.15}$$

It seems that we are stuck here. The integration by parts does not seem to help much, as it only transfers $\int e^x \sin x \, dx$ to $\int e^x \cos x \, dx$, which is not an improvement. In this case, we have to pause and think, what happens if we use the integration by parts again for $\int e^x \cos x \, dx$? We let $u = e^x$ and $dv/dx = \cos x$, which gives $du/dx = e^x$ and $v = \sin x$, and we have

$$\int e^x \cos x \, dx = e^x \sin x - \int \sin x \frac{du}{dx} dx = e^x \sin x - \int e^x \sin x \, dx. \tag{7.16}$$

Then, you may think, what is the use, as we have come back to the original integral? But what happens if we combine the above two equations? We have

$$I = \int e^x \sin x \, dx = -e^x \cos x - \int e^x \cos x \, dx$$

$$= -e^x \cos x + e^x \sin x - \int e^x \sin x = e^x(\sin x - \cos x) - I. \tag{7.17}$$

We want to find I, so we can solve the above equation by simple rearrangement

$$2I = e^x(\sin x - \cos x), \tag{7.18}$$

or

$$I = \frac{1}{2}e^x(\sin x - \cos x). \tag{7.19}$$

Here we have found the integral as

$$I = \int e^x \sin x dx = \frac{1}{2}e^x(\sin x - \cos x) + C. \tag{7.20}$$

Such a procedure is very useful in deriving reduction formulae for some integrals. Let us take the Wallis' formula as an example.

Example 7.4:   In order to derive Wallis' formula for the integral

$$I_n = \int_0^{\pi/2} \sin^n x dx, \qquad (n \geq 2). \tag{7.21}$$

We use $u = \sin^{n-1} x$ and $dv/dx = \sin x$; we get

$$\frac{du}{dx} = (n-1)\sin^{n-2} x \cos x, \qquad v = \int \sin x = -\cos x.$$

We have

$$I_n = \left[-\sin^{n-1} x \cos x \right]_0^{\pi/2} - \int_0^{\pi/2} (-\cos x)(n-1)\sin^{n-2} x \cos x dx$$

$$= 0 + (n-1) \int_0^{\pi/2} \cos^2 x \sin^{n-2} dx.$$

Since $\sin^2 x + \cos^2 x = 1$ or $\cos^2 x = 1 - \sin^2 x$, we have

$$I_n = (n-1) \int_0^{\pi/2} (1 - \sin^2 x)\sin^{n-2} dx$$

$$= (n-1) \int_0^{\pi/2} \sin^{n-2} dx - (n-1) \int_0^{\pi/2} \sin^n x dx = (n-1)I_{n-2} - (n-1)I_n,$$

whose rearrangement leads to

$$I_n = \frac{n-1}{n}I_{n-2},$$

which is Wallis' reduction formula for the integral I_n. If we continue to use the reduction formula, when n is odd, we have

$$I_n = \frac{(n-1)(n-3)...4 \times 2}{n(n-2)...5 \times 3}.$$

When n is even, we have

$$I_n = \frac{(n-1)(n-3)...3 \times 1}{n(n-2)...4 \times 2} \cdot \frac{\pi}{2}.$$

It is left as an exercise to prove that the integral $\int_0^{\pi/2} \cos^n x dx$ follows the same formulae.

When using integration by parts, sometimes we have to try to use different combinations for u and v. In the following example, we have to use $u = x$ and $v' = xe^{-x^2}$. If we try to use $u = x^2$, it will not work.

Example 7.5: Let us now try to calculate the integral

$$J = \int_{-\infty}^{\infty} x^2 e^{-x^2} dx,$$

using integration by parts or $\int u dv = uv - \int v du$. By setting $u = x$ and $v' = xe^{-x^2}$, we have

$$\frac{du}{dx} = 1, \qquad v = \int xe^{-x^2} dx = \frac{1}{2} \int e^{-x^2} d(x^2) = -\frac{1}{2} e^{-x^2}.$$

We have

$$J = \int_{-\infty}^{\infty} x^2 e^{-x^2} dx = \left[-\frac{x}{2} e^{-x^2} \right]_{-\infty}^{\infty} - \int_{-\infty}^{\infty} (\frac{-1}{2} e^{-x^2}) dx$$

$$= 0 + \frac{1}{2} \int_{-\infty}^{\infty} e^{-x^2} dx = \frac{\sqrt{\pi}}{4} \left[\mathrm{erf}(x) \right]_{-\infty}^{\infty} = \frac{\sqrt{\pi}}{4} \times [1 - (-1)] = \frac{\sqrt{\pi}}{2},$$

where we have used the error function

$$\mathrm{erf}(x) = \frac{2}{\sqrt{\pi}} \int_0^x e^{-u^2} du,$$

and $\mathrm{erf}(\infty) = 1$ and $\mathrm{erf}(-\infty) = -1$. We will use this result in deriving the variance of the Gaussian distribution.

7.3 Integration by Substitution

Sometimes, it is not easy to carry out the integration directly. However, it might become easier if we use the change of variables or integration by substitution. For example, we want to calculate the integral

$$I = \int f(x) dx. \tag{7.22}$$

We can change the variable x into another variable $u = g(x)$ where $g(x)$ is a known function of x. This means

$$\frac{du}{dx} = g'(x), \qquad (7.23)$$

or

$$du = g'(x)dx, \qquad dx = \frac{1}{g'(x)}du. \qquad (7.24)$$

This means that

$$I = \int f(x)dx = \int f[g^{-1}(u)]\frac{1}{g'}du, \qquad (7.25)$$

where it is usually not necessary to calculate $g^{-1}(u)$ as it is relatively obvious. For example, in order to do the integration

$$I = \int xe^{x^2+5}dx, \qquad (7.26)$$

we let $u = x^2 + 5$ and we have

$$\frac{du}{dx} = (x^2 + 5)' = 2x, \qquad (7.27)$$

or

$$du = 2xdx, \qquad (7.28)$$

which means $dx = \frac{1}{2x}du$. Therefore, we have

$$I = \int xe^{x^2+5}dx = \int xe^u \frac{1}{2x}du$$

$$= \frac{1}{2}\int e^u du = \frac{1}{2}e^u + A = e^{x^2+5} + A, \qquad (7.29)$$

where A is the constant of integration. Here we have substituted $u = x^2 + 5$ back in the last step.

Example 7.6: Compaction occurs when loosely packaged materials (such as soils and snow) deform under their own weight, resulting in the reduction of porosity. Compacted sediments will form sedimentary rocks, and compacted snow will form glaciers. In the simplest case, the porosity ϕ will decrease almost exponentially in terms of

$$\phi = \phi_0 e^{-\alpha z},$$

where z is the depth of the column of the porous materials, and ϕ_0 is the initial porosity at $z = 0$. The parameter α can be considered as a constant, typically $\alpha = 0.001/\text{m}$. In sedimentary basins, the pores can be considered fully saturated with fluid, and the reduction of pore space is often linked

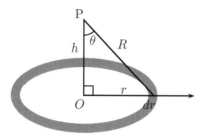

Figure 7.4: Derivation of the Bouguer correction formula.

with the generation of hydrocarbon under the right conditions. Suppose a fraction (say, $\beta = 0.01$ or about 1%) of such fluids has been turned into oil, then the total volume can be estimated by the total reduction of the pore space

$$V = A\beta \int_0^h (\phi_0 - \phi)dz = \beta A \int_0^h \phi_0(1 - e^{-\alpha z})dz,$$

where A is the area of interest. Setting $u = \alpha z$, we have

$$V = \beta A \phi_0 \int_0^h dz - \beta A \phi_0 \int_0^{\alpha h} \frac{1}{\alpha} e^{-u} du = \beta A \phi_0 h - \frac{\beta A}{\alpha} \phi_0 [e^{-\alpha h} - 1].$$

When changing variables, the integration limits should also change accordingly. Otherwise, the results will be different and incorrect. This will be demonstrated by the application in the next section.

7.4   Bouguer Gravity

The Bouguer formula for gravity correction is derived using an infinite plate of thickness t with a uniform density ρ. As the plate is infinite, we can choose any point as the origin $r = 0$ and cut a small ring, and we are only interested in the gravity at point P outside the plate at a perpendicular distance of h away (see Fig. 7.4).

Due to symmetry, the gravitational force will point towards point O. According to Newton's law of gravitation, the gravity caused by the small ring is

$$dg = -\frac{Gdm}{R^2} \cos\theta, \tag{7.30}$$

where G is the universal gravitational constant, θ is the angle. R is the distance from P to the ring. The mass of the ring is $dm = \rho t dA$ and

$dA = 2\pi r dr$ is the area of the ring with the thickness of dr. Thus, we have

$$dm = 2\pi \rho t r dr. \tag{7.31}$$

In addition, from the basic trigonometry, we know that

$$\cos\theta = \frac{h}{R}, \qquad R = \sqrt{h^2 + r^2}. \tag{7.32}$$

Therefore, the gravity due to the ring becomes

$$dg = -2\pi G \rho h t \frac{r dr}{R^3} = -2\pi G \rho h t \frac{r dr}{(h^2 + r^2)^{3/2}}. \tag{7.33}$$

So the total gravity due to the whole plate is the integration

$$\Delta g = -2\pi G \rho h t \int_0^\infty \frac{r dr}{(h^2 + r^2)^{3/2}}. \tag{7.34}$$

Using the change of variables by setting $u = (r^2 + h^2)$ so that $du = 2rdr$ and $u = h^2$ for $r = 0$, we have

$$\Delta g = -2\pi G \rho h t \int_{h^2}^\infty \frac{1}{2} \frac{du}{u^{3/2}}. \tag{7.35}$$

Since $\int u^{-3/2} du = -2u^{-1/2}$, we have

$$\Delta g = -2\pi G \rho h t \int_{h^2}^\infty \left[-\frac{1}{u^{1/2}} \right]_{h^2}^\infty$$

$$= -2\pi G \rho h t [-(0 - \frac{1}{h})] = -2\pi G \rho t. \tag{7.36}$$

This is exactly the well-known formula for Bouguer reduction of a layer with a density of ρ and thickness t.

Using the typical values of $G = 6.67 \times 10^{-11}$ m^3/s^2 kg, and density $\rho = 2670$ kg/m^2 for rocks, we have

$$\Delta g = -2\pi \times 6.67 \times 10^{-11} \times 2670 \, t$$

$$\approx 1.1189 \times 10^{-6} t \ (\text{m/s}^2) \approx 0.112 \, t \ (\text{mGal}). \tag{7.37}$$

This is to say, for every 1-metre layer removed, the Bouguer gravity anomaly is corrected by -0.112 mGal. The Bouguer gravity variations reflect changes in mass distribution below the surface after removing most of the effect of mass excess above the reference level, often the sea level.

Chapter 8

Fourier Transforms

Mathematical transform is a method of changing one kind of function and equation into another, often simpler and easier to solve. Fourier transform maps a function in the time domain such as a signal into another function in the frequency domain, which is commonly used in signal processing. In earth sciences, the processing of seismic signals is pivotal to oil and gas exploration and the understanding of almost all underground geological structures.

As an application, we will analyse the Milankovitch cycles using simple Fourier transform in the last section of this chapter.

8.1 Fourier Series

From earlier discussions, we know that the function e^x can be expanded into a series in terms of a polynomial with terms of x, x^2, ..., x^n. In this case, we are in fact trying to expand it in terms of the basis functions 1, x, x^2, ..., and x^n. There are many other basis functions. For example, the basis functions $\sin(n\pi t)$ and $\cos(n\pi t)$ are more widely used in seismic signal processing. In general, this is essentially about the Fourier series. For a function $f(t)$ on an interval $t \in [-T, T]$ where $T > 0$ is a finite constant or half period, the Fourier series is defined as

$$f(t) = \frac{a_0}{2} + \sum_{n=1}^{\infty} \left[a_n \cos(\frac{n\pi t}{T}) + b_n \sin(\frac{n\pi t}{T}) \right], \tag{8.1}$$

where

$$a_0 = \frac{1}{T} \int_{-T}^{T} f(t) dt, \qquad a_n = \frac{1}{T} \int_{-T}^{T} f(t) \cos(\frac{n\pi t}{T}) dt, \tag{8.2}$$

and

$$b_n = \frac{1}{T} \int_{-T}^{T} f(t) \sin(\frac{n\pi t}{T}) dt, \quad (n = 1, 2, ...). \tag{8.3}$$

Here a_n and b_n are the Fourier coefficients of $f(t)$ on $[-T, T]$. The function $f(t)$ can be continuous or piecewise continuous with a finite number of jump discontinuities. For a jump discontinuity at $t = t_0$, if $f'(t_0-)$ and $f'(t_0+)$ both exist with $f(t_0-) \neq f(t_0+)$, then the Fourier series converge to the average value. That is

$$f(t_0) = \frac{1}{2}[f(t_0-) + f(t_0+)]. \tag{8.4}$$

You may wonder how to calculate the coefficient a_n and b_n? Before we proceed, let us prove the orthogonality relation

$$J = \int_{-T}^{T} \sin(\frac{n\pi t}{T}) \sin(\frac{m\pi t}{T}) dt = \begin{cases} 0 & (n \neq m) \\ T & (n = m) \end{cases}, \tag{8.5}$$

where n and m are integers.

From the trigonometrical functions, we know that

$$\cos(A + B) = \cos A \cos B - \sin A \sin B, \tag{8.6}$$

and

$$\cos(A - B) = \cos A \cos B + \sin A \sin B. \tag{8.7}$$

By subtracting, we have

$$\cos(A - B) - \cos(A + B) = 2 \sin A \sin B. \tag{8.8}$$

Now the orthogonality integral becomes

$$J = \int_{-T}^{T} \sin(\frac{n\pi t}{T}) \sin(\frac{m\pi t}{T}) dt$$

$$= \frac{1}{2} \int_{-T}^{T} \left\{ \cos[\frac{(n-m)\pi t}{T}] - \cos[\frac{(n+m)\pi t}{T}] \right\} dt. \tag{8.9}$$

If $n \neq m$, we have

$$J = \frac{1}{2} \left\{ \frac{T}{(n-m)\pi} \sin[\frac{(n-m)\pi t}{T}] \Big|_{-T}^{T} - \frac{T}{(n+m)\pi} \sin[\frac{(n+m)\pi t}{T}] \Big|_{-T}^{T} \right\}$$

$$= \frac{1}{2}[\frac{T}{(n-m)\pi} \times (0 - 0) - \frac{T}{(n+m)\pi} \times (0 - 0)] = 0. \tag{8.10}$$

If $n = m$, we have

$$J = \frac{1}{2} \int_{-T}^{T} \left\{ 1 - \cos[\frac{2n\pi t}{T}] \right\} dt$$

$$= \frac{1}{2}\left\{t\Big|_{-T}^{T} - \frac{T}{2n\pi}\sin[\frac{2n\pi t}{T}]\Big|_{-T}^{T}\right\} = \frac{1}{2}[2T - \frac{T}{2n\pi} \times 0] = T, \quad (8.11)$$

which proves the relation (8.5). Using similar calculations, we can easily prove the following orthogonality relations

$$\int_{-T}^{T} \cos(\frac{n\pi t}{T})\cos(\frac{m\pi t}{T})dt = \begin{cases} 0 & (n \neq m) \\ T & (n = m) \end{cases}, \quad (8.12)$$

and

$$\int_{-T}^{T} \sin(\frac{n\pi t}{T})\cos(\frac{m\pi t}{T})dt = 0, \qquad \text{for all } n \text{ and } m. \quad (8.13)$$

Now we can try to derive the expression for the coefficients a_n. Multiplying both sides of the Fourier series (8.1) by $\cos(m\pi t/T)$ and taking the integration from $-T$ to T, we have

$$\int_{-T}^{T} f(t)\cos(\frac{m\pi t}{T})dt = \frac{a_0}{2}\int_{-T}^{T}\cos(\frac{m\pi t}{T})dt$$

$$+ \sum_{n=1}^{\infty}\left\{a_n\int_{-T}^{T}\cos(\frac{n\pi t}{T})\cos(\frac{m\pi t}{T})dt + b_n\int_{-T}^{T}\sin(\frac{n\pi t}{T})\cos(\frac{m\pi t}{T})dt\right\}.$$

Using the relations (8.12) and (8.13) as well as $\int_{-T}^{T}\cos(m\pi t/T)dt = 0$, we know that the only non-zero integral on the right-hand side is when $n = m$. Therefore, we get

$$\int_{-T}^{T} f(t)\cos(\frac{n\pi t}{T})dt = 0 + [a_n T + b_n \times 0], \quad (8.14)$$

which gives

$$a_n = \frac{1}{T}\int_{-T}^{T} f(t)\cos(\frac{n\pi t}{T})dt, \quad (8.15)$$

where $n = 1, 2, 3, \cdots$. Interestingly, when $n = 0$, it is still valid and becomes a_0 as $\cos 0 = 1$. That is

$$a_0 = \frac{1}{T}\int_{-T}^{T} f(t)dt. \quad (8.16)$$

In fact, $a_0/2$ is the average of $f(t)$ over the period $2T$. The coefficients b_n can be obtained by multiplying $\sin(m\pi t/T)$ and following similar calculations.

Fourier series in general tends to converge slowly. In order for a function $f(x)$ to be expanded, it must satisfy the Dirichlet conditions: $f(x)$ must be periodic with at most a finite number of discontinuities,

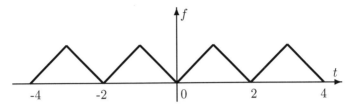

Figure 8.1: Triangular wave with a period of 2.

and/or a finite number of minima or maxima within one period. In addition, the integral of $|f(x)|$ must converge. For example, these conditions suggest that $\ln(x)$ cannot be expanded into a Fourier series in the interval $[0, 1]$ as $\int_0^1 |\ln x| dx$ diverges.

The nth term of the Fourier series,

$$a_n \cos(n\pi t/T) + b_n \sin(n\pi t/T),$$

is called the nth harmonic. The energy of the nth harmonic is defined by $A_n^2 = a_n^2 + b_n^2$, and the sequence of A_n^2 forms the energy or power spectrum of the Fourier series.

From the coefficient a_n and b_n, we can easily see that $b_n = 0$ for an even function $f(-t) = f(t)$ because $g(t) = f(t) \sin(n\pi t/T)$ is now an odd function $g(-t) = -g(t)$ due to the fact $\sin(2\pi t/T)$ is an odd function. We have

$$b_n = \frac{1}{T} \int_{-T}^{T} f(t) \sin(\frac{n\pi t}{T}) dt = \frac{1}{T} \left[\int_{-T}^{0} g(t)dt + \int_{0}^{T} g(t)dt \right]$$

$$= \frac{1}{T} \left[\int_{0}^{T} g(-t)dt + \int_{0}^{T} g(t)dt \right] = \frac{1}{T} \int_{0}^{T} (-g(t) + g(t))dt = 0. \quad (8.17)$$

Similarly, we have $a_0 = a_n = 0$ for an odd function $f(-t) = -f(t)$. In both cases, only one side $[0, T]$ of the integration is used due to symmetry. Thus, for even function $f(t)$, we have the Fourier cosine series on $[0,T]$

$$f(t) = \frac{a_0}{2} + \sum_{n=1}^{\infty} a_n \cos(\frac{n\pi t}{T}). \quad (8.18)$$

For odd function $f(t)$, we have the sine series $f(t) = \sum_{n=1}^{\infty} \sin(\frac{n\pi t}{T})$.

Example 8.1: The triangular wave is defined by $f(t) = |t|$ for $t \in [-1, 1]$ with a period of 2 or $f(t + 2) = f(t)$ shown in Fig. 8.1. Using the coefficients of the Fourier series, we have

$$a_0 = \int_{-1}^{1} |t| dt = \int_{-1}^{0} (-t)dt + \int_{0}^{1} t dt = 1.$$

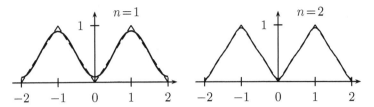

Figure 8.2: Fourier series for the triangular wave $f(t) = |t|, t \in [-1, 1]$:
(a) first two terms $(n=1)$; (b) first three terms $(n = 2)$.

Since both $|t|$ and $\cos(n\pi t)$ are even functions, we have for any $n \geq 1$,

$$a_n = \int_{-1}^{1} |t| \cos(n\pi t) dt = 2 \int_{0}^{1} t \cos(n\pi t) dt$$

$$= 2 \frac{t}{n\pi} \sin(n\pi t) \Big|_{0}^{1} - \frac{2}{n\pi} \int_{0}^{1} \sin(n\pi t) dt = \frac{2}{n^2\pi^2} [\cos(n\pi) - 1].$$

Because $|t| \sin(n\pi t)$ is an odd function, we have

$$b_n = \int_{-1}^{1} |t| \sin(n\pi t) dt = 0.$$

Hence, the Fourier series for the triangular wave can be written as

$$f(t) = \frac{1}{2} + 2 \sum_{n=1}^{\infty} \frac{\cos(n\pi) - 1}{n^2\pi^2} \cos(n\pi t) = \frac{1}{2} + \frac{4}{\pi^2} \sum_{n=1,3,5,\dots}^{\infty} \frac{(-1)^n}{n^2} \cos(n\pi t).$$

The first few terms, $f_n(t) = 1/2 + 4/\pi^2 \cos(\pi t)$, are shown in Fig. 8.2 where we can see that only a few terms are needed to produce a very good approximation.

Here we can see that the triangular wave with derivative disconti-nuity can be approximated well by two or three terms. This makes it easy for any mathematical analysis. Fourier series are widely applied in signal processing.

8.2 Fourier Transforms

In general, when the period T becomes infinite, the Fourier coefficients of a function defined on the whole real axis $(-\infty, \infty)$ can be written as

$$a(\omega_n) = \int_{-T}^{T} f(t) \cos(\omega_n t) dt, \qquad b(\omega_n) = \int_{-T}^{T} f(t) \sin(\omega_n t) dt, \quad (8.19)$$

where $\omega_n = \frac{n\pi}{T}$ under the limits of $T \to \infty$ and $\omega_n \to 0$. If we further pose the constraint $\int_{-\infty}^{\infty} |f(t)| < \infty$, we get $a_0 \to 0$. In this case, the Fourier series becomes the Fourier integral

$$f(t) = \int_0^\infty [a(\omega)\cos(\omega t) + b(\omega)\sin(\omega t)]d\omega, \qquad (8.20)$$

where

$$a(\omega) = \frac{1}{\pi} \int_{-\infty}^\infty f(t)\cos(\omega t)dt, \; b(\omega) = \frac{1}{\pi} \int_{-\infty}^\infty f(t)\sin(\omega t)dt. \quad (8.21)$$

Following similar discussions above, even functions lead to Fourier cosine integrals and odd functions lead to Fourier sine integrals.

The Fourier transform $\mathcal{F}[f(t)]$ of $f(t)$ is defined as

$$F(\omega) = \mathcal{F}[f(t)] = \frac{1}{\sqrt{2\pi}} \int_{-\infty}^\infty f(t)e^{-i\omega t}dt, \qquad (8.22)$$

and the inverse Fourier transform can be written as

$$f(t) = \mathcal{F}^{-1}[F(\omega)] = \frac{1}{\sqrt{2\pi}} \int_{-\infty}^\infty F(\omega)e^{i\omega t}d\omega, \qquad (8.23)$$

where $\exp[i\omega t] = \cos(\omega t) + i\sin(\omega t)$. The Fourier transform has the following properties:

$$\mathcal{F}[f(t) + g(t)] = \mathcal{F}[f(t)] + \mathcal{F}[g(t)], \qquad \mathcal{F}[\alpha f(t)] = \alpha \mathcal{F}[f(t)], \quad (8.24)$$

and $\mathcal{F}[(-it)^n f(t)] = \frac{d^n F(\omega)}{d\omega^n}$. There are some variations of the transforms such as the Fourier sine transform and the Fourier cosine transform. The Fourier transforms of some common functions are listed in Table 8.1.

8.3   DFT and FFT

Now we try to write the Fourier series (8.1) in a complex form using $\cos\theta = (e^{i\theta} + e^{-i\theta})/2$ and $\sin\theta = (e^{i\theta} - e^{-i\theta})/2i$, we have the nth term

$$f_n(t) = a_n \cos(\frac{n\pi t}{T}) + b_n \sin(\frac{n\pi t}{T}) = \frac{a_n[e^{\frac{in\pi t}{T}} + e^{\frac{-in\pi t}{T}}]}{2} + \frac{b_n[e^{\frac{in\pi t}{T}} - e^{\frac{-in\pi t}{T}}]}{2i}$$

$$= \frac{(a_n - ib_n)}{2}e^{in\pi t/T} + \frac{(a_n + ib_n)}{2}e^{-in\pi t/T}. \qquad (8.25)$$

If we define $\beta_n = \frac{(a_n - ib_n)}{2}$, and $\beta_{-n} = \frac{(a_n + ib_n)}{2}$ where $(n = 0, 1, 2, ...)$, and set $\beta_0 = a_0/2$, we get $f_n(t) = \beta_n e^{-in\pi t/T} + \beta_{-n}e^{-in\pi t/T}$. Therefore, the Fourier series can be written in the complex form

$$f(t) = \sum_{n=-\infty}^\infty \beta_n e^{i\pi nt/T}. \qquad (8.26)$$

Table 8.1: Fourier Transforms

f(t)	$F(\omega) = \mathcal{F}[f(t)]$
$f(t - t_0)$	$F(\omega)e^{-i\omega t_0}$
$f(t)e^{-i\omega_0 t}$	$F(\omega - \omega_0)$
$\delta(t)$	$1/\sqrt{2\pi}$
1	$\sqrt{2\pi}\delta(\omega)$
$e^{-(\alpha t)^2} \quad (\alpha > 0)$	$\frac{1}{\sqrt{2}\alpha}e^{-\frac{\omega^2}{4\alpha^2}}$
$\frac{1}{\alpha^2 + t^2}$	$\sqrt{\frac{\pi}{2}}\frac{e^{-\alpha|\omega|}}{\alpha}$
$\cos(\omega_0 t)$	$\sqrt{\frac{\pi}{2}}[\delta(\omega - \omega_0) + \delta(\omega + \omega_0)]$
$\sin(\omega_0)$	$i\sqrt{\frac{\pi}{2}}[\delta(\omega + \omega_0) - \delta(\omega - \omega_0)]$
$\frac{\sin \alpha x}{x} \quad (\alpha > 0)$	$\sqrt{\frac{\pi}{2}}, \quad (|\omega| < \alpha); 0, \quad (|\omega| > \alpha)$

In signal processing, we are often interested in $f(t)$ in $[0, 2T]$ (rather than $[-T, T]$). Without loss of generality, we can set $T = \pi$. In this case, the Fourier coefficients become

$$\beta_n = \frac{1}{2\pi} \int_0^{2\pi} f(t)e^{-int} dt. \tag{8.27}$$

As it is not easy to compute these coefficients accurately, we often use the numerical integration to approximate the above integral with a step size $h = 2\pi/N$. This is equivalent to sampling t with N sample points $t_k = 2\pi k/N$ where $k = 0, 1, ..., N - 1$. Therefore, the coefficient can be estimated by $\beta_n = \frac{1}{2\pi} \int_0^{2\pi} f(t)e^{-int} dt \approx \frac{1}{N} \sum_{k=0}^{N-1} f(\frac{2\pi k}{N})e^{-2\pi ink/N}$. Once we know β_n, we know the whole spectrum. Let f_k denote $f(2\pi k/N)$, we can define the discrete Fourier transform (DFT) as

$$F_n = \sum_{k=0}^{N-1} f_k e^{-\frac{2\pi nk}{N}}, \tag{8.28}$$

which is for periodic discrete signals $f(k)$ with a period of N. A periodic signal $f(k + N) = f(k)$ has a periodic spectrum $F(n + N) = F(n)$. The discrete Fourier transform consists of N multiplications and $N - 1$ additions for each F_n, thus for N values of n, the computational complexity is of $O(N^2)$.

In fact, by rearranging the formulae, we can get a class of fast Fourier transform (FFT) whose computational complexity is about $O(N \log(N))$. Using the notation $\omega = e^{-2\pi i/N}$ and $\omega^N = e^{2\pi i} = 1$, we can rewrite (8.28) as

$$F_n = \sum_{k=0}^{N-1} f_k \omega^{kn}, \quad -\infty < n < \infty. \tag{8.29}$$

It is worth pointing out that $\omega = e^{-2\pi i/N}$ is the Nth root of unity, thus the powers of ω always lie on a unit circle in the complex plane. Here, the computations only involve the summation and the power of ω.

This usually requires a lot of computations; however, in the case when N can be factorised, some of the calculations can be decomposed into different steps and many of calculations become unnecessary. In this case, we often use $N = 2^m$ where m is a positive integer; it becomes the so-called FFT, and the computational complexity is now reduced to $2N \log_2(N)$. For example, when $N = 2^{20}$, FFT will reduce the computational time from three weeks to less than a minute on modern desktop computers. There is a huge amount of literature about FFT, filter design, signal reconstruction and their applications in seismic signal processing.

8.4   Milankovitch Cycles

Now let us look at a real-world example by studying the Milankovitch cycles in climate changes. Milankovitch theory explains paleoclimate fluctuations and occurrence of the Ice Ages very well. The Milankovitch cycles, named after the scientist M. Milankovitch who studied the effect of the Earth's orbital motion on the climate in a pioneer paper published in 1941, refer to the collective effect on climate change due to the changes in the Earth's orbital movements (see Fig. 8.3).

There are three major components in the orbital changes: precession of the perihelion, obliquity (or wobble of the Earth's axis of rotation), and eccentricity (or shape of the Earth's orbit). Because of the interaction of the Sun, the Moon, and other planets (mainly Jupiter and Saturn) with the Earth, each of the three components usually has multiple harmonic components. Here we will outline the theory.

The precession of the perihelion has a number of harmonic components, ranging from 19 to 23.7 thousand years (kyrs), though the weighted averaged is about 21 kyrs. The tilting of the Earth's axis of rotation varies from about 21.5° to 24.5° with periods from 29 to 53.6 kyrs. The averaged period is about 41.6 kyrs. The increase of obliquity will lead to the increase of the amplitude of the seasonal cycle in insolation. At the same time, the precession or wobble of this axis (relative to fixed stars) completes a big circle in about 26 kyrs, though it is about 21 kyrs if calculated relative to the perihelion. This wobble is mainly caused by the differential gravitational force due to the fact that the Earth is not a perfect sphere and it has an equatorial bulge.

The change of eccentricity varies from $e = 0.005$ to 0.06 with periods ranging from 94.9 to 412.9 kyrs. Two major components are a long period of 412.9 kyrs and a short (average) period of 110.7 kyrs, and the latter is close to the 100 kyrs cycles of ice ages. All these harmonic

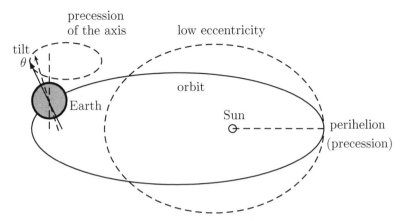

Figure 8.3: Milankovitch cycles of the Earth's orbital elements.

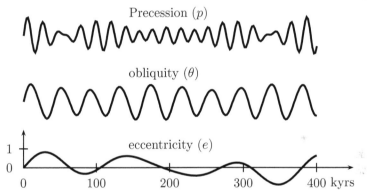

Figure 8.4: Milankovitch cycles: (a) precession of perihelion;
(b) obliquity; and (c) eccentricity.

components interact and result in a complicated climate pattern.

From Berger's calculations in 1977 based on Milankovitch's theory, we can write the precession as

$$p \approx p_0 + \tilde{p}\left[0.42\sin(\frac{2\pi t}{19.1}) + 0.28\sin(\frac{2\pi t}{22.4}) + 0.30\sin(\frac{2\pi t}{23.7})\right], \quad (8.30)$$

where we have used the approximated averaged periods. $\tilde{p}$ is the averaged amplitude of the precession and p_0 is the initial value. In writing this equation, we have implicitly assumed that the phase shift between different harmonic components is negligible, and components with similar periods have been combined into a single major component.

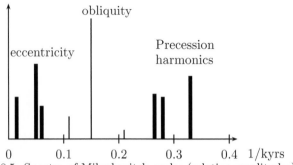

Figure 8.5: Spectra of Milankovitch cycles (relative amplitudes).

Similarly, the obliquity can be expressed as

$$\theta \approx \theta_0 + \tilde{\theta}\left[0.06\sin(\frac{2\pi t}{29}) + 0.80\sin(\frac{2\pi t}{41}) + 0.14\sin(\frac{2\pi t}{53.6})\right], \quad (8.31)$$

where $\tilde{\theta} \approx 1.5°$ is the averaged amplitude of tilting and $\theta_0 \approx 23°$ is the mean angle. The current tilting is about 23.44°.

The variation of the eccentricity is

$$e \approx e_0 + \tilde{e}\left[0.22\sin(\frac{2\pi t}{95}) + 0.50\sin(\frac{2\pi t}{125}) + 0.28\sin(\frac{2\pi t}{412.9})\right], \quad (8.32)$$

where $\tilde{e} \approx 0.0275$ is the averaged amplitude of eccentricity, and $e_0 \approx 0.0325$ is the mean eccentricity. The present eccentricity of the Earth's orbit is about 0.017. Although the variation of e is small, it still results in a change of distance of the order of about 5 million kilometres (aphelion minus perihelion), or about 3% of the average distance from the Earth to the Sun, which will result in about 6% change in solar energy reaching the Earth as the energy flux is inversely proportional to the distance.

These variations are shown in Fig. 8.4 and their amplitude spectra are shown in Fig. 8.5.

Chapter 9

Vectors

The quantities or variables we have discussed so far are scalar as we have been concerned with the magnitude only. However, in many cases, we have to consider the direction as well. For example, driving a car travelling south at a speed of 30 mph is very different from travelling north at the same speed, because we are now dealing with not only the magnitude, but also the direction. A quantity with magnitude and direction is called a vector, and many quantities such as velocity, force, displacement and acceleration are vectors. In this chapter, we will first introduce vectors and vector products.

Later in this chapter, we will demonstrate how to use vectors in modelling real-world problems such as the Mohr-Coulomb failure criterion, thrust faults, and electrical prospecting.

9.1 Vector Algebra

9.1.1 Vectors

Suppose we travel from a point P at (x_1, y_1) to another point Q, (say, a weather station) at (x_2, y_2), we have a displacement vector $\boldsymbol{d} = \overrightarrow{PQ}$ (see Fig. 9.1).

For the displacement vector, we need the magnitude (or the length or distance) between P and Q, and also the direction or angle θ to determine the vector uniquely. Since the coordinates of two points P and Q are given, the distance between P and Q can be calculated using the Cartesian distance. The length or magnitude of $\boldsymbol{d}$ can conveniently be written as $d = |\boldsymbol{d}|$, and we have

$$PQ = |\overrightarrow{PQ}| = |\boldsymbol{d}| = \sqrt{(x_2 - x_1)^2 + (y_2 - y_1)^2}. \qquad (9.1)$$

Here we follow the conventions of using a single letter in italic form to denote the magnitude while using the same letter in bold type to

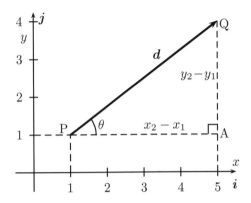

Figure 9.1: The displacement from $P(x_1, y_1)$ to point $Q(x_2, y_2)$.

denote the vector itself.

The direction of the vector is represented by the angle θ from the x-axis. We have

$$\tan \theta = \frac{y_2 - y_1}{x_2 - x_1}. \tag{9.2}$$

Conventionally, we often write a vector using bold font d, rather than d. In many books, vectors are also written in the overhead arrow form such as $\overrightarrow{PQ}$ or simply $\vec{d}$. The notation $\overrightarrow{PQ}$ signifies that the vector is pointing from P to Q. Here we will use the bold-type notations as they are more popularly used in mathematics. The components of the vector d are $x_2 - x_1$ along the x-axis and $y_2 - y_1$ along the y-axis. This provides a way to write the vector as

$$d = \overrightarrow{PQ} = \begin{pmatrix} x_2 - x_1 \\ y_2 - y_1 \end{pmatrix}. \tag{9.3}$$

Here we write the vector as a column, called a column vector.

Example 9.1: Reading from the graph shown in Fig. 9.1, we know that P is at $(1, 1)$ and Q is at $(5, 4)$. The displacement can be represented in mathematical form

$$d = \begin{pmatrix} x_2 - x_1 \\ y_2 - y_1 \end{pmatrix} = \begin{pmatrix} 5 - 1 \\ 4 - 1 \end{pmatrix} = \begin{pmatrix} 4 \\ 3 \end{pmatrix}.$$

Therefore, the distance PQ or the magnitude d of the displacement d is

$$d = |d| = \sqrt{(5-1)^2 + (4-1)^2} = \sqrt{4^2 + 3^2} = 5.$$

The angle θ is given by

$$\tan \theta = \frac{4 - 1}{5 - 1} = \frac{3}{4} = 0.75,$$

or

$$\theta = \tan^{-1} = 0.75 \approx 36.87°.$$

For real-world problems, the unit of length must be given, either in km or metres or any other suitable units.

- -

It is worth pointing out that the vector $\overrightarrow{QP}$ is pointing the opposite direction $\overrightarrow{PQ}$, and we thus have

$$\overrightarrow{QP} = \begin{pmatrix} x_1 - x_2 \\ y_1 - y_2 \end{pmatrix} = \begin{pmatrix} (-1)(x_2 - x_1) \\ (-1)(y_2 - y_1) \end{pmatrix} = -\begin{pmatrix} x_2 - x_1 \\ y_2 - y_1 \end{pmatrix} = -\overrightarrow{PQ} = -\boldsymbol{d}.$$

In general for any real number $\beta \neq 0$ and a vector $\boldsymbol{v} = \begin{pmatrix} a \\ b \end{pmatrix}$, we have

$$\beta \boldsymbol{v} = \beta \begin{pmatrix} a \\ b \end{pmatrix} = \begin{pmatrix} \beta a \\ \beta b \end{pmatrix}. \tag{9.4}$$

A vector whose magnitude is 1 is called a unit vector. So all the following vectors are unit vectors

$$\boldsymbol{i} = \begin{pmatrix} 1 \\ 0 \end{pmatrix}, \quad \boldsymbol{j} = \begin{pmatrix} 0 \\ 1 \end{pmatrix}, \quad \boldsymbol{w} = \begin{pmatrix} \cos \theta \\ \sin \theta \end{pmatrix}. \tag{9.5}$$

For any θ, we know $|\boldsymbol{w}| = \sqrt{\cos^2 \theta + \sin^2 \theta} = 1$ due to the identity $\sin^2 \theta + \cos^2 \theta = 1$. The vectors $\boldsymbol{i}$ and $\boldsymbol{j}$ are the unit vectors along x-axis and y-axis directions, respectively.

Since a vector has a magnitude and a direction, any two vectors with the same magnitude and direction should be equal since there is no other constraint. This means that we can shift and move both ends of a vector by any same amount in any direction, and we still have the same vector. In other words, if two vectors are equal, they must have the same magnitude and direction. Mathematically, their corresponding components must be equal.

If no physical barrier is our concern, then we can reach point Q from point P in an infinite number of ways. We can go along the x-axis direction to the right for a distance $x_2 - x_1$ to the point A, and then go upward along the y-direction for a distance $y_2 - y_1$. This is equivalent to saying that $\boldsymbol{d}$ is the sum of two vectors $(x_2 - x_1)\boldsymbol{i}$ and $(y_2 - y_1)\boldsymbol{j}$. We have

$$\boldsymbol{d} = \begin{pmatrix} x_2 - x_1 \\ y_2 - y_1 \end{pmatrix} = (x_2 - x_1)\boldsymbol{i} + (y_2 - y_1)\boldsymbol{j}. \tag{9.6}$$

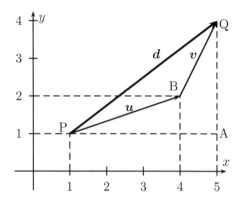

Figure 9.2: The displacement from point P(1,1) to point Q(5,4).

Similarly, we can first go along the direction of PB and then along BQ (see Fig. 9.2). This also suggests that

$$\vec{PQ} = \vec{PB} + \vec{BQ}, \quad \text{or} \quad \boldsymbol{d} = \boldsymbol{u} + \boldsymbol{v}. \tag{9.7}$$

Now the point B is at $(4, 2)$. So we have

$$\boldsymbol{u} = \begin{pmatrix} 4-1 \\ 2-1 \end{pmatrix} = \begin{pmatrix} 3 \\ 1 \end{pmatrix}, \quad \boldsymbol{v} = \begin{pmatrix} 5-4 \\ 4-2 \end{pmatrix} = \begin{pmatrix} 1 \\ 2 \end{pmatrix}, \quad \boldsymbol{d} = \begin{pmatrix} 4 \\ 3 \end{pmatrix}, \tag{9.8}$$

which suggests that

$$\boldsymbol{u} + \boldsymbol{v} = \begin{pmatrix} 3 \\ 1 \end{pmatrix} + \begin{pmatrix} 1 \\ 2 \end{pmatrix} = \begin{pmatrix} 3+1 \\ 1+2 \end{pmatrix} = \begin{pmatrix} 4 \\ 3 \end{pmatrix} = \boldsymbol{d}. \tag{9.9}$$

The addition of two vectors is a vector whose components are simply the addition of their corresponding components. If we define the subtraction of any two vectors $\boldsymbol{u}$ and $\boldsymbol{v}$ as

$$\boldsymbol{u} - \boldsymbol{v} = \boldsymbol{u} + (-\boldsymbol{v}), \tag{9.10}$$

where $-\boldsymbol{v}$ is obtained by flipping $\boldsymbol{v}$ by $180°$. In general, we have

$$\boldsymbol{v}_1 \pm \boldsymbol{v}_2 = \begin{pmatrix} a_1 \\ b_1 \end{pmatrix} \pm \begin{pmatrix} a_2 \\ b_2 \end{pmatrix} = \begin{pmatrix} a_1 \pm a_2 \\ b_1 \pm b_2 \end{pmatrix}. \tag{9.11}$$

The addition of any two vectors $\boldsymbol{u}$ and $\boldsymbol{v}$ is commutative, that is

$$\boldsymbol{v}_1 + \boldsymbol{v}_2 = \boldsymbol{v}_2 + \boldsymbol{v}_1. \tag{9.12}$$

This is because each of its components is commutative: $a_1 + a_2 = a_2 + a_1$ and $b_1 + b_2 = b_2 + b_1$. Similarly, as the addition of scalars is associative

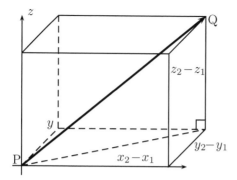

Figure 9.3: The 3D displacement vector from $P(x_1, y_1, z_1)$ to point $Q(x_2, y_2, z_2)$.

(i.e., $a_1 + (a_2 + a_3) = (a_1 + a_2) + a_3$), then the addition of vectors is associative as well. That is

$$v_1 + (v_2 + v_3) = (v_1 + v_2) + v_3. \tag{9.13}$$

So far we have only focused on the vectors in a two-dimensional plane; we can easily extend our discussion to 3D vectors or higher-dimensional vectors. For the 3D vector shown in Fig. 9.3, we have

$$d = \overrightarrow{PQ} = \begin{pmatrix} x_2 - x_1 \\ y_2 - y_1 \\ z_2 - z_1 \end{pmatrix}. \tag{9.14}$$

If we define the unit vectors as

$$i = \begin{pmatrix} 1 \\ 0 \\ 0 \end{pmatrix}, \qquad j = \begin{pmatrix} 0 \\ 1 \\ 0 \end{pmatrix}, \qquad k = \begin{pmatrix} 0 \\ 0 \\ 1 \end{pmatrix}, \tag{9.15}$$

for the three perpendicular directions, we can write d as

$$d = \begin{pmatrix} x_2 - x_1 \\ y_2 - y_1 \\ z_2 - z_1 \end{pmatrix} = (x_2 - x_1)i + (y_2 - y_1)j + (z_2 - z_1)k. \tag{9.16}$$

The addition and subtraction of any two vectors now becomes

$$v_1 \pm v_2 = \begin{pmatrix} a_1 \\ b_1 \\ c_1 \end{pmatrix} \pm \begin{pmatrix} a_2 \\ b_2 \\ c_2 \end{pmatrix} = \begin{pmatrix} a_1 \pm a_2 \\ b_1 \pm b_2 \\ c_1 \pm c_2 \end{pmatrix}. \tag{9.17}$$

These formulae can be extended to the addition and subtraction of multiple vectors.

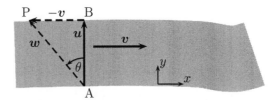

Figure 9.4: Swimming across a river with the water velocity v. The swimmer must swim upstream along w so as to get to B.

Example 9.2: A swimmer intends to swim across a river at a constant speed 2 m/s from point A to B. The water in the river flows at an average velocity $v = 1$m/s, and the river is $d = 25$ metres wide. He or she has to aim at a different angle θ along $w = \overrightarrow{AP}$ at a speed 2 m/s, rather than directly at B (see Fig. 9.4), otherwise he or she will reach some point downstream. The vector w is determined by

$$w = u - v = u + (-v), \quad \text{or} \quad u = w + v.$$

In order to calculate the angle θ, we can simply use the trigonometrical functions for the triangle $\triangle ABP$. That is

$$\sin \theta = 1/2 = 0.5, \quad \text{or} \quad \theta = \sin^{-1} 0.5 \approx 30°.$$

So the velocity along y-axis swimming across the river is

$$u = w \cos \theta = 2 \cos 30° = 1.73 \text{ m/s}.$$

The time taken to cross the river is $t = \frac{d}{u} = \frac{25}{1.73} = 14.4$ seconds.

It is worth pointing out that if $|u| < |v|$, there is no solution. That is to say, it is impossible to reach point B from A, and the swimmer will reach somewhere downstream.

9.1.2 Product of Vectors

We have just discussed the addition and subtraction of vectors. A natural question is whether we can construct any multiplication and division. There are different ways to carry out multiplication of vectors, but the division of vectors does not have any meaningful applications in earth sciences.

The product of two vectors can be either a scalar or a vector, depending on the way we carry out the multiplications. The scalar product of two vectors F and d is defined as

$$F \cdot d = Fd \cos \theta. \tag{9.18}$$

This rather odd definition has some physical meaning. We know that the work W done by a force f to move an object a distance s, is simply $W = fs$ on the condition that the force is applied along the direction of movement. If a force $\mathbf{F}$ is applied at an angle θ related to the displacement $\mathbf{d}$ (see Fig. 9.5), we first have to decompose or project the force $\mathbf{F}$ onto the displacement direction so that the component actually acts on the object along the direction of $\mathbf{d}$ is $\mathbf{F}_{\parallel} = F \cos\theta$. So the actual work done becomes

$$W = \mathbf{F}_{\parallel} d = Fd \cos\theta, \tag{9.19}$$

which means that the amount of work W is the scalar product

$$W = \mathbf{F} \cdot \mathbf{d}. \tag{9.20}$$

Here the $\cdot$ symbol denotes such a scalar product. From such notations, the scalar product of two vectors is also called the dot product or inner product.

If we intend to compute in terms of their components

$$\mathbf{F} = \begin{pmatrix} f_1 \\ f_2 \\ f_3 \end{pmatrix}, \qquad \mathbf{d} = \begin{pmatrix} d_1 \\ d_2 \\ d_3 \end{pmatrix}, \tag{9.21}$$

the dot product can be calculated by

$$\mathbf{F} \cdot \mathbf{d} = f_1 d_1 + f_2 d_2 + f_3 d_3. \tag{9.22}$$

Since $\cos 90° = 0$, when the scalar product is zero, it suggests that the two vectors are perpendicular to each other; sometimes we also say they are orthogonal. So for the unit vectors $\mathbf{i}, \mathbf{j}$, and $\mathbf{k}$, we have

$$\mathbf{i} \cdot \mathbf{i} = \mathbf{j} \cdot \mathbf{j} = \mathbf{k} \cdot \mathbf{k} = 1, \quad \text{or} \quad \mathbf{i} \cdot \mathbf{j} = \mathbf{j} \cdot \mathbf{k} = \mathbf{k} \cdot \mathbf{i} = 0. \tag{9.23}$$

These basic properties can easily be verified by using the formula (9.22).

If we know the dot product, we can use it to determine the angle θ, and we have

$$\cos\theta = \frac{\mathbf{F} \cdot \mathbf{d}}{Fd}. \tag{9.24}$$

The dot product has some interesting properties. From its definition, it is easy to see that $\mathbf{F} \cdot \mathbf{d} = \mathbf{d} \cdot \mathbf{F}$. Another interesting property is the distributive law:

$$\mathbf{F} \cdot (\mathbf{d} + \mathbf{s}) = \mathbf{F} \cdot \mathbf{d} + \mathbf{F} \cdot \mathbf{s}. \tag{9.25}$$

Now let us prove the above distributive law. Using

$$\mathbf{s} = \begin{pmatrix} s_1 \\ s_2 \\ s_3 \end{pmatrix}, \tag{9.26}$$

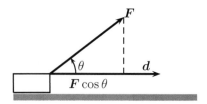

Figure 9.5: Work done W by a force F to move an object in the direction of displacement d is $W = F \cdot d = Fd \cos \theta$.

we have

$$F \cdot d = f_1 d_1 + f_2 d_2 + f_3 d_3, \qquad F \cdot s = f_1 s_1 + f_2 s_2 + f_3 s_3. \qquad (9.27)$$

Now we have

$$F \cdot (d + s) = F \cdot \left[\begin{pmatrix} d_1 \\ d_2 \\ d_3 \end{pmatrix} + \begin{pmatrix} s_1 \\ s_2 \\ s_3 \end{pmatrix} \right] = F \cdot \begin{pmatrix} d_1 + s_1 \\ d_2 + s_2 \\ d_3 + s_3 \end{pmatrix}$$

$$= f_1(d_1 + s_1) + f_2(d_2 + s_2) + f_3(d_3 + s_3)$$

$$= (f_1 d_1 + f_2 d_2 + f_3 d_3) + (f_1 s_1 + f_2 s_2 + f_3 s_3) = F \cdot d + F \cdot s, \quad (9.28)$$

which is the distributive law.

The vector product, also called the cross product or outer product, of two vectors u and v forms another vector w. The definition can be written as

$$u \times v = uv \sin \theta \, n, \qquad (9.29)$$

where n is the unit vector, and w points the direction of n which is perpendicular to both vectors u and v, forming a right-handed system (see Fig. 9.6). In addition, θ is the angle between u and v, and u and v are the magnitudes of u and v, respectively.

In many books, the notation $u \wedge v$ is also used, that is

$$u \wedge v \equiv u \times v. \qquad (9.30)$$

The right-handed system suggests that, if we change the order of the product, there is a sign change. That is $v \times u = -u \times v$.

Though the vector product is a vector; however, its magnitude has a geometrical meaning. That is, the magnitude is the area of the shaded parallelogram shown in Fig. 9.6.

Using $u = \begin{pmatrix} u_1 & u_2 & u_3 \end{pmatrix}^T$, and $v = \begin{pmatrix} v_1 & v_2 & v_3 \end{pmatrix}^T$ where the superscript T means the transpose which turns a column vector into a

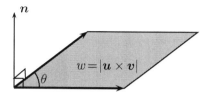

Figure 9.6: The direction of $u \times v$ points the direction along n while the magnitude $w = |u \times v| = |u||v| \sin \theta$ is the area of the shaded region.

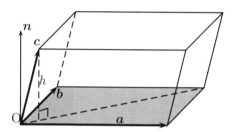

Figure 9.7: The triple product of three vectors a, b, and c and the volume of the parallelepiped.

row vector or vice versa, we can write the vector product in terms of their components

$$u \times v = \begin{pmatrix} u_2 v_3 - u_3 v_2 \\ u_3 v_1 - u_1 v_3 \\ u_1 v_2 - u_2 v_1 \end{pmatrix}$$

$$= (u_2 v_3 - u_3 v_2)i + (u_3 v_1 - u_1 v_3)j + (u_1 v_2 - u_2 v_1)k. \qquad (9.31)$$

For any three vectors a, b, and c, their combination is not always meaningful. For example, $\beta = a \cdot b$ gives a scalar, as we need two vectors to form a dot product, therefore, the combination of $c \cdot (a \cdot b)$ is meaningless, as is $(a \cdot b) \cdot c$. However, the combination $(a \cdot b)c$ is meaningful if we interpret it as $(a \cdot b)c = \beta c$.

For three vectors, we can define a scalar triple product that is widely used in vector analysis with geometrical interpretation. The scalar triple product of three vectors is defined by

$$V = c \cdot (a \times b), \qquad (9.32)$$

which is the volume of the parallelepiped formed by the three vectors (see Fig. 9.7). We know that the vector product $S = a \times b = Sn$ (where $S = ab \sin \theta$ and n is the unit vector) is pointing the direction of n, perpendicular to both a and b. The magnitude S of S is the area of the base parallelogram formed by a and b. Then, the dot product of c and the unit vector n gives the high h as it essentially projects the vector c onto the unit vector n. So the scalar triple product $c \cdot (Sn)$ now becomes the product of the base area and the perpendicular height h, which is exactly the volume of the parallelepiped.

Using the notations of the components

$$a = \begin{pmatrix} a_1 \\ a_2 \\ a_3 \end{pmatrix}, \qquad b = \begin{pmatrix} b_1 \\ b_2 \\ b_3 \end{pmatrix}, \qquad c = \begin{pmatrix} c_1 \\ c_2 \\ c_3 \end{pmatrix}, \qquad (9.33)$$

we have

$$c \cdot (a \times b) = (a_2 b_3 - a_3 b_2)c_1 + (a_3 b_1 - a_1 b_3)c_2 + (a_1 b_2 - a_2 b_1)c_3. \quad (9.34)$$

This provides a way to calculate the scalar triple product using their components.

Example 9.3: For three vectors,

$$a = \begin{pmatrix} 4 \\ 1 \\ 0 \end{pmatrix}, \qquad b = \begin{pmatrix} 0 \\ 2 \\ -1 \end{pmatrix}, \qquad c = \begin{pmatrix} -1 \\ 2 \\ 5 \end{pmatrix},$$

we can calculate the volume of the parallelepiped formed by these three vectors using the scalar triple product $V = c \cdot (a \times b)$. Since

$$S = a \times b = \begin{pmatrix} a_2 b_3 - a_3 b_2 \\ a_3 b_1 - a_1 b_3 \\ a_1 b_2 - a_2 b_1 \end{pmatrix} = \begin{pmatrix} 1 \times (-1) - 0 \times 2 \\ 0 \times 0 - 4 \times (-1) \\ 4 \times 2 - 1 \times 0 \end{pmatrix} = \begin{pmatrix} -1 \\ 4 \\ 8 \end{pmatrix}.$$

So the scalar triple product or the volume becomes

$$V = c \cdot S = -1 \times (-1) + 2 \times 4 + 5 \times 8 = 49.$$

9.2   Gradient and Laplace Operators

For a given scalar function $\psi(x, y, z)$, the gradient vector is defined by

$$\text{grad}\psi \equiv \nabla\psi = \frac{\partial \psi}{\partial x}i + \frac{\partial \psi}{\partial y}j + \frac{\partial \psi}{\partial z}k, \qquad (9.35)$$

where i, j, k are the unit vectors along x-, y- and z-directions, respectively. The notation grad or ∇ are interchangeable. Here ∇ is the gradient operator defined as

$$\nabla = \frac{\partial}{\partial x}i + \frac{\partial}{\partial y}j + \frac{\partial}{\partial z}k, \qquad (9.36)$$

which is commonly known as the 'del' operator. For example, if we know the distribution of the temperature T of the Earth, we can calculate the heat flux q as $q = -K\nabla T$, where K is the heat conductivity. In the simplest one-dimensional case when we are only concerned with the temperature variation with depth z, we have $q = -K\partial T/\partial z$.

Example 9.4: Darcy's law for flow in porous media can be written as

$$q = -\frac{k}{\mu}\nabla p,$$

where ∇p is the pressure gradient, k is the permeability of the media, and μ is the viscosity of water. q is called the Darcy flux or velocity which is related to the pore velocity or seepage velocity v by $q = v\phi$ where ϕ is the porosity. In hydrology, Darcy's law is often expressed in terms of the water head h as

$$q = -K\nabla h,$$

where K is the hydraulic conductivity, and ∇h is called the hydraulic gradient. Suppose in an aquifer, we have $\nabla h = 0.5$ and $K = 4 \times 10^{-5}$ m/s, then the Darcy velocity of the groundwater flow is approximately

$$q = -4 \times 10^{-5} \times 0.5 = -2 \times 10^{-5} \text{ m/s} \approx -1.7 \text{ m/day}.$$

It would take about a month to flow through a layer 50 metres thick.

Let us look at another example.

Example 9.5: In the calculation of gravity variations, it is often easy to calculate the gravitational potential V first, then the force can be obtained by taking its gradient. For two masses M and m with a distance r apart, the gravitational potential energy is

$$V(r) = -\frac{GMm}{r}, \qquad (9.37)$$

where G is the universal gravitational constant.

The force between the two objects is

$$F = -\nabla V = GMm\nabla\left(\frac{1}{r}\right) = -\frac{GMm}{r^2}n = -\frac{GMmr}{|r|^3}, \qquad (9.38)$$

where $n = r/|r|$ is the unit vector along r. This is essentially the vector form of Newton's law of gravitation. It is worth pointing out that in some books $F = \nabla V$ is used, and this depends on the convention whether the force of attraction is defined as positive or negative.

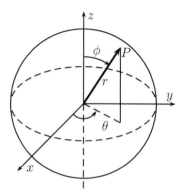

Figure 9.8: Spherical coordinates (r, θ, ψ) for any $P(x, y, z)$.

Another related operator is the divergence operator for a vector field $\boldsymbol{f}(x, y, z) = (u, v, w)^T$ where u, v and w are its three components. The divergence of $\boldsymbol{f}$ is defined as

$$\text{div} \boldsymbol{f} \equiv \boldsymbol{\nabla} \cdot \boldsymbol{f} = \frac{\partial u}{\partial x} + \frac{\partial v}{\partial y} + \frac{\partial w}{\partial z}, \tag{9.39}$$

which is a scalar.

The proper combination of the grad with div will lead to the Laplace operator ∇^2 for a scalar function ψ

$$\Delta \psi \equiv \nabla^2 \psi \equiv \boldsymbol{\nabla} \cdot (\boldsymbol{\nabla} \psi) = \frac{\partial^2 \psi}{\partial x^2} + \frac{\partial^2 \psi}{\partial y^2} + \frac{\partial^2 \psi}{\partial z^2}. \tag{9.40}$$

The famous Laplace equation can be written as

$$\nabla^2 \psi = \frac{\partial^2 \psi}{\partial x^2} + \frac{\partial^2 \psi}{\partial y^2} + \frac{\partial^2 \psi}{\partial z^2} = 0, \tag{9.41}$$

which is important in applications, and its solutions are related to harmonic functions, such as the free oscillations of the Earth, and the harmonic expansions of the Earth's gravity and the geodesy.

The spherical coordinates are most widely used in earth sciences, though the mathematical definition is slightly different from the latitude and longitude system for the Earth. In the three-dimensional case, the spherical coordinates, also called spherical polar coordinates, (r, θ, ϕ) are shown in Fig. 9.8. The angle θ is the azimuthal angle in the $x - y$ plane, and $0 \leq \theta \leq 2\pi$. It is similar to the longitude but with a different range. The polar angel angle ϕ is the angle from z-axis, and typically $0 \leq \psi \pi$. Latitude λ is related to ϕ by $\lambda = 90° - \phi$.

For any point $P(x, y, z)$, it is relatively straightforward to derive the relationship between x, y, z and r, θ, ψ using trigonometry. For example, $z = r \cos \phi$. The relationships can be written as

$$x = r \sin \phi \cos \theta, \qquad y = r \sin \phi \sin \theta, \qquad z = r \cos \phi. \qquad (9.42)$$

In the spherical coordinates, the gradient and Laplace operators become

$$\nabla V = \frac{\partial V}{\partial r} \boldsymbol{u}_r + \frac{1}{r \sin \phi} \frac{\partial V}{\partial \theta} \boldsymbol{u}_\theta + \frac{1}{r} \frac{\partial V}{\partial \phi} \boldsymbol{u}_\phi, \qquad (9.43)$$

where $\boldsymbol{u}_r, \boldsymbol{u}_\theta$ and $\boldsymbol{u}_\phi$ are the unit vectors along r, θ and ϕ directions, respectively. In the simplest case when the function ψ does not depend on θ and ψ, the gradient is simply $\nabla V = \partial V/\partial r$ along the direction of $\boldsymbol{u}_r$. We will use this result in the application of the electrical method in geophysical prospecting.

The Laplace operator in the spherical coordinates can be written as

$$\nabla^2 \psi = \frac{1}{r^2} \frac{\partial}{\partial r} \left(r^2 \frac{\partial \psi}{\partial r} \right) + \frac{1}{r^2 \sin \theta} \frac{\partial}{\partial \theta} \left(\sin \theta \frac{\partial \psi}{\partial \theta} \right) + \frac{1}{r^2 \sin \theta^2} \frac{\partial^2 \psi}{\partial \phi^2}. \qquad (9.44)$$

Interested readers can refer to more advanced literature for detailed derivations.

9.3 Applications

9.3.1 Mohr-Coulomb Criterion

Let us see a dry block resting on a slope inclined at an angle θ to the horizontal direction (see Fig. 9.9). The weight of the block is $W = mg$ where m is its mass and g is the acceleration due to gravity. The driving force F_d down the slope is the weight vector projecting on the direction along the slip surface. That is $F_d = W \sin \theta$.

The normal force component perpendicular to the slip surface is $F_n = W \cos \theta$, which means that the friction force F_μ is determined by

$$F_\mu = \mu F_n = \mu W \cos \theta, \qquad (9.45)$$

where μ is the friction coefficient. At the steady state, the block is just able to slip with an almost uniform velocity, requiring that $F_\mu - F_d = 0$ or the net force is zero. We have

$$\mu W \cos \theta - W \sin \theta = 0, \qquad (9.46)$$

or

$$\mu = \frac{\sin \theta}{\cos \theta} = \tan \theta. \qquad (9.47)$$

The block slips down if the downward force is greater than the friction force. Alternatively, we can consider the situation when the angle θ is adjustable gradually from $\theta = 0$ (horizontal) to a steep angle. When θ

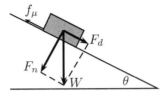

Figure 9.9: Friction coefficient.

reaches a critical angle $\theta = \phi$ such that the block is just able to slip, ϕ is the angle of friction.

In most applications, we are more concerned with the shear stress τ and the normal stress σ. In order for a block to slip, the shear stress along the slip surface must be at least equal to $\sigma \tan \phi$. That is $\tau = \sigma \tan \phi$.

In reality, there is some cohesion between the two contact surfaces (similar to putting some glue between the surfaces). Let S be the cohesion per unit area. The block will only slip if the shear stress is greater than the combined resistance of the friction stress $\sigma \tan \phi$ and the cohesion. That is

$$\tau \geq S + \sigma \tan \phi. \tag{9.48}$$

The equality is the well-known Mohr-Coulomb yield criterion for the failure in soil and porous materials.

$$\tau_* = S + \sigma \tan \phi, \tag{9.49}$$

where τ_* is the critical shear stress or failure shear stress. In this case, the angle ϕ is called the angle of friction or friction angle. The detailed derivations of this criterion require the use of stress tensor and Mohr's circle; however, the introduction here is a very crude way to show how it works. For example, if the block is a soil block (or a fault), the slip movement will cause cracks and failure in soils (or crust). The failure is called a shear failure, and the slip plane is called the failure plane. Failure will occur if the slope θ is steeper than ϕ, i.e., $\theta > \phi$.

The cohesion stress is usually very small compared with the normal stress level. For example, for limestones, we have $\tau_* = 10 + 0.85\sigma$(MPa), which suggests a friction angle $\phi = \tan^{-1} 0.85 \approx 40°$. For most granular materials and in most applications in earth sciences, $S = 0$ is a good approximation. So we have $\tau = \sigma \tan \phi$. In the case of a wet block or the presence of pore fluids or water, we have a modified criterion

$$\tau = (\sigma - p) \tan \phi = (1 - \lambda)\sigma \tan \phi, \tag{9.50}$$

where p is the pore pressure and $\lambda = p/\sigma$ is a ratio describing the effect of pore pressure on the failure.

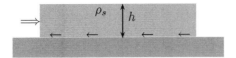

Figure 9.10: Thrust fault paradox.

9.3.2 Thrust Faults

The thrust fault paradox states that neither gravitational nor tectonic forces are sufficient to push large thrust fault sheets over a long distance. Either the stress required to move a thrust sheet is much higher than the strength of the rock, or the slope angle is too high. This paradox can be illustrated by considering the thrust sheet as a horizontal block with a thickness h overlaying and sliding over another fixed block (see Fig. 9.10). The stress at the interface is $\sigma_n = \rho g h$, where ρ is the density of the thrust sheet and g is the acceleration due to gravity. To move the block against friction, we have to overcome the frictional resistance $\sigma_h = f_h$

$$\sigma_h = \mu \sigma_n = \mu \rho g h = \rho g (\tan \phi) h, \tag{9.51}$$

where $\mu = \tan \phi$ is the friction coefficient.

The rock failure stress is about 250 MPa= 2.5×10^8 Pa. For $\mu = 0.577$, we have the thickness

$$h = \frac{\sigma_h}{\mu \rho g} \approx \frac{2.5 \times 10^8}{0.577 \times 2600 \times 9.8} \approx 17005 \text{ m} \approx 17 \text{ km}. \tag{9.52}$$

This seems to imply that $h = 17$ km. Alternatively, we can view this as pushing a block of maximum length of 17 km from behind. But the actual thrust fault could be a few kilometers thick, up to 150 km wide and tens of kilometres.

If the sheet is driven by its weight, it has to be a slope with an angle θ so that $\tan \theta > \tan \phi$ or

$$\rho g h \tan \theta \geq \rho g h \tan \phi. \tag{9.53}$$

Experiments suggest that $\mu \approx 0.577$ or $\phi \approx 30°$. This means that $\theta > 30°$ and this value is too high for most real thrust faults. In reality, the pore fluid at the interface may be very important, which will effectively lower the effective stress and reduce the frictional force. This can be modelled by $\sigma_h = \mu(1 - \lambda)\rho g h$, where $\lambda = 0.4$ to 1 is the pore fluid factor, defined as the ratio of pore pressure to lithostatic pressure. Obviously, $\lambda \to 1$, then frictional resistance is virtually zero.

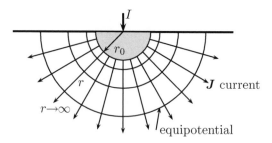

Figure 9.11: Electrical potential for a single electrode.

9.3.3 Electrical Method in Prospecting

The porous materials below the Earth's surface can act as a conductive medium. If we know the electrical resistance R, we can calculate the current I for any given electrical potential V using Ohm's law

$$V = IR, \qquad \text{or} \quad I = \frac{V}{R}. \tag{9.54}$$

From basic physics, we know how to calculate R for a thin wire with a length of L and a cross section area A. That is $R = \frac{\gamma L}{A}$ whose unit is Ω. γ is the electrical resistivity in the unit of Ω m. The common notation of resistivity is ρ, but we use γ in this book to avoid any possible confusion with the density notation ρ.

The reciprocal of γ is often called electrical conductivity σ. That is $\sigma = 1/\gamma$, which has a unit of siemens per metre or S/m. The more general form of Ohm's law can be written as the gradient

$$\boldsymbol{J} = -\sigma \boldsymbol{\nabla} V = -\frac{1}{\gamma} \boldsymbol{\nabla} V, \tag{9.55}$$

where $\boldsymbol{J}$ is the current density or the current per unit area, which is in fact a vector as it has a magnitude and a direction.

In an idealised case when there is only one electrode of radius r_0 (see Fig. 9.11), the other electrode is at infinity so that the voltage is zero. How do we relate the input current I with the voltage V? Though we are dealing with a 3D shell, this is essentially a one-dimensional problem in terms of radial distance r due to symmetry.

So for a shell at r with a thickness dr, the increment of voltage dV can be related to the current I by

$$dV = I\frac{\gamma dr}{A} = I\gamma \frac{dr}{2\pi r^2}, \tag{9.56}$$

where we have used the area of the shell (half-sphere) as $2\pi r^2$.

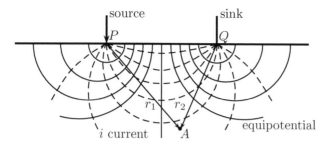

Figure 9.12: Equipotential curves and current flow around two electrodes in a homogeneous media.

Integrating the above equation with respect to r from $r = r_0$ to ∞ and using the boundary condition $V \to 0$ as $r \to \infty$, we have

$$V = \int_{r_0}^{\infty} dv = \frac{I\gamma}{2\pi} \int_{r_0}^{\infty} \frac{1}{r^2} dr = \frac{I\gamma}{2\pi} \left[\frac{-1}{r} \right]_{r_0}^{\infty} = \frac{I\gamma}{2\pi r_0}. \tag{9.57}$$

This is a very simple but important relationship for electrical methods in geophysical prospecting.

In the more common case when two electrodes are presented, one is called a source and the other a sink (see Fig. 9.12). The potential V_1 of the source at any point A with a distance r_1 and the potential V_2 of the sink at A are

$$V_1 = \frac{I\gamma}{2\pi r_1}, \qquad V_2 = \frac{-I\gamma}{2\pi r_2}, \tag{9.58}$$

where the negative sign comes from the fact that the current is flowing out from the sink. Since the potential is a scalar, we can simply superpose their potentials, and we have

$$V_A = V_1 + V_2 = \frac{I\gamma}{2\pi}\left(\frac{1}{r_1} - \frac{1}{r_2}\right). \tag{9.59}$$

For a given value of $V_a = V = \text{const}$, the combination of r_1 and r_2 will trace out a family of curves, called equipotential lines or curves (actually surfaces in 3D). They are shown in Fig. 9.12 as the circular heavy curves. In the same figure, the dashed curves are the paths along which the local current flows, which can be obtained by taking the gradient of V. Now we have the current density

$$\boldsymbol{J} = -\frac{1}{\gamma}\boldsymbol{\nabla}V = -\frac{1}{\gamma}\frac{I\gamma}{2\pi}\boldsymbol{\nabla}[\frac{1}{r_1} - \frac{1}{r_2}] = -\frac{I}{2\pi}\boldsymbol{\nabla}[\frac{1}{r_1} - \frac{1}{r_2}], \tag{9.60}$$

which can be expressed in terms of the coordinates (x, y) at A, though the detailed calculations would be tedious.

However, in practice, we are more interested in the inverse problem as the media is inhomogeneous, and the resistivity γ varies with locations. Since we know the input current I, and we can measure the voltage at different locations, the main problem is to try to estimate the resistivity γ and its relationship with underground geological structures. As the presence of fluids can affect the conductivity significantly, it is therefore widely used in groundwater survey and oil prospecting.

Chapter 10

Matrix Algebra

The vector concept and algebra we have just discussed can be extended to matrices. In fact, a vector is a very special class of matrices. In this chapter, we will introduce the fundamentals of matrices and linear algebra. As an application, we will use the eigenvalue technique to discuss the natural frequencies of mechanical vibrations.

10.1 Matrices

A matrix is a rectangular array of numbers. For example, a coffee shop sells four different type of coffees, and the sales in terms of the numbers of cups for three consecutive days are recorded as follows:

$$
\begin{array}{ccccc}
 & A & B & C & D \\
\text{Day 1} & 210 & 256 & 197 & 207 \\
\text{Day 2} & 242 & 250 & 205 & 199 \\
\text{Day 3} & 192 & 249 & 220 & 215
\end{array}
\tag{10.1}
$$

where different products form a row, and different days for the same product form a column. This can be written as the sale matrix with 3 rows and 4 columns.

$$
S = \begin{pmatrix} 210 & 256 & 197 & 207 \\ 242 & 250 & 205 & 199 \\ 192 & 249 & 220 & 215 \end{pmatrix}.
\tag{10.2}
$$

Each item of the numbers is called an entry or element of the matrix. We usually use a bold-type upper case to denote a matrix, and we use the lower case to denote its elements. Therefore, we have

$$
S = [s_{ij}], \qquad (i = 1, 2, 3, \text{ and } j = 1, 2, 3, 4).
\tag{10.3}
$$

The element on the second row and the third column is $s_{23} = 205$.

The transpose of a matrix S can be obtained by interchanging its rows and columns, and is denoted by S^T. We have

$$S^T = \begin{pmatrix} 210 & 256 & 197 & 207 \\ 242 & 250 & 205 & 199 \\ 192 & 249 & 220 & 215 \end{pmatrix}^T = \begin{pmatrix} 210 & 242 & 192 \\ 256 & 250 & 249 \\ 197 & 205 & 220 \\ 207 & 199 & 215 \end{pmatrix}. \tag{10.4}$$

The same coffee shop owner owns another coffee shop on a different street, selling the same products. The same three-day sales are

$$Q = \begin{pmatrix} 191 & 229 & 170 & 240 \\ 195 & 209 & 199 & 214 \\ 207 & 272 & 149 & 190 \end{pmatrix}. \tag{10.5}$$

The total sales of both shops are obtained by the addition of their corresponding entries

$$S + Q = \begin{pmatrix} 210 & 256 & 197 & 207 \\ 242 & 250 & 205 & 199 \\ 192 & 249 & 220 & 215 \end{pmatrix} + \begin{pmatrix} 191 & 229 & 170 & 240 \\ 195 & 209 & 199 & 214 \\ 207 & 272 & 149 & 190 \end{pmatrix}$$

$$= \begin{pmatrix} 210+191 & 256+229 & 197+170 & 207+240 \\ 242+195 & 250+209 & 205+199 & 199+214 \\ 192+207 & 249+272 & 220+149 & 215+190 \end{pmatrix}$$

$$= \begin{pmatrix} 401 & 485 & 367 & 447 \\ 437 & 459 & 404 & 413 \\ 399 & 521 & 369 & 405 \end{pmatrix}. \tag{10.6}$$

Their sales differences are

$$S - Q = \begin{pmatrix} 210 & 256 & 197 & 207 \\ 242 & 250 & 205 & 199 \\ 192 & 249 & 220 & 215 \end{pmatrix} - \begin{pmatrix} 191 & 229 & 170 & 240 \\ 195 & 209 & 199 & 214 \\ 207 & 272 & 149 & 190 \end{pmatrix}$$

$$= \begin{pmatrix} 210-191 & 256-229 & 197-170 & 207-240 \\ 242-195 & 250-209 & 205-199 & 199-214 \\ 192-207 & 249-272 & 220-149 & 215-190 \end{pmatrix}$$

$$= \begin{pmatrix} 19 & 27 & 27 & -33 \\ 47 & 41 & 6 & -15 \\ -15 & -23 & 71 & 25 \end{pmatrix}. \tag{10.7}$$

We can see here that the addition and subtraction of the matrices are carried out entry by entry. It is only possible to carry out addition and subtraction if and only if the matrices S and Q have the same numbers of rows and columns.

The prices for each product are: GBP 0.99 for A, GBP 1.50 for B, GBP 1.15 for C, and GBP 0.90 for D. This can be written as a column matrix or a column vector

$$p = \begin{pmatrix} 0.99 \\ 1.50 \\ 1.15 \\ 0.90 \end{pmatrix}. \tag{10.8}$$

The total sales income for each day is given by the multiplication of S and p.

$$I_1 = Sp = \begin{pmatrix} 210 & 256 & 197 & 207 \\ 242 & 250 & 205 & 199 \\ 192 & 249 & 220 & 215 \end{pmatrix} \begin{pmatrix} 0.99 \\ 1.50 \\ 1.15 \\ 0.90 \end{pmatrix} \tag{10.9}$$

$$= \begin{pmatrix} 210 \times 0.99 + 256 \times 1.50 + 197 \times 1.15 + 207 \times 0.90 \\ 242 \times 0.99 + 250 \times 1.50 + 205 \times 1.15 + 199 \times 0.90 \\ 192 \times 0.99 + 249 \times 1.50 + 220 \times 1.15 + 215 \times 0.90 \end{pmatrix} = \begin{pmatrix} 1004.75 \\ 1029.43 \\ 1010.08 \end{pmatrix}.$$

which means that the total incomes for the three days are GBP 1004.75, GBP 1029.43 and GBP 1010.08, respectively. In general, the multiplication of two matrices is possible if and only if the number of columns of the first matrix on the left (S) is the same as the number of rows of the second matrix on the right p. If $A = [a_{ij}]$ is an $m \times n$ matrix, and $B = [b_{jk}]$ is an $n \times p$ matrix, then $C = AB$ is an $m \times p$ matrix. We have

$$C = [c_{ik}] = AB = [a_{ij}][b_{jk}], \qquad c_{ik} = \sum_{j=1}^{n} a_{ij} b_{jk}. \tag{10.10}$$

When a scalar α multiplies a matrix A, the result is the matrix with each of A's elements multiplying by α. For example,

$$\alpha A = \alpha \begin{pmatrix} a & b \\ c & d \\ e & f \end{pmatrix} = \begin{pmatrix} \alpha a & \alpha b \\ \alpha c & \alpha d \\ \alpha e & \alpha f \end{pmatrix}. \tag{10.11}$$

Similarly, the sales incomes at another shop are given by

$$I_2 = Qp = \begin{pmatrix} 191 & 229 & 170 & 240 \\ 195 & 209 & 199 & 214 \\ 207 & 272 & 149 & 190 \end{pmatrix} \begin{pmatrix} 0.99 \\ 1.50 \\ 1.15 \\ 0.90 \end{pmatrix} = \begin{pmatrix} 944.09 \\ 928.00 \\ 955.28 \end{pmatrix}. \tag{10.12}$$

So the total incomes for the same owner at both shops are

$$I_T = I_1 + I_2 = \begin{pmatrix} 1004.75 \\ 1029.43 \\ 1010.08 \end{pmatrix} + \begin{pmatrix} 944.09 \\ 928.00 \\ 955.28 \end{pmatrix} = \begin{pmatrix} 1948.84 \\ 1957.43 \\ 1965.36 \end{pmatrix}. \tag{10.13}$$

Generally speaking, the addition of two matrices is commutative

$$S + Q = Q + S. \tag{10.14}$$

The addition of three matrices is associative, that is

$$(S + Q) + A = S + (Q + A). \tag{10.15}$$

However, matrices multiplication is not commutative. That is

$$AB \neq BA. \tag{10.16}$$

There are two special matrices: the zero matrix and the identity matrix. A zero matrix is a matrix whose every element is zero. We have 1×4 zero matrix as $O = \begin{pmatrix} 0 & 0 & 0 & 0 \end{pmatrix}$. If the number of rows of a matrix is the same as the number columns, that is $m = n$, the matrix is called a square matrix. If the diagonal elements are 1's and all the other elements are zeros, it is called a unit matrix or an identity matrix. For example, the 3×3 unit matrix can be written as

$$I = \begin{pmatrix} 1 & 0 & 0 \\ 0 & 1 & 0 \\ 0 & 0 & 1 \end{pmatrix}. \tag{10.17}$$

If a matrix A is the same size as the unit matrix, then it is commutative. That is

$$IA = AI = A. \tag{10.18}$$

10.2   Transformation and Inverse

When a point $P(x, y)$ is rotated by an angle θ, it becomes its corresponding point $P'(x', y')$ (see Fig. 10.1). The relationship between the old coordinates (x, y) and the new coordinates (x', y') can be derived by using the trigonometry.

The new coordinates at the new location $P'(x', y')$ can be written in terms of the coordinates at the original location $P(x, y)$ and the angle θ. From basic geometry, we know that $\angle A'P'Q = \theta = \angle A'OW$ Since $x = OA = OA'$, $y = AP = A'P'$, and $x' = OS = OW - SW = OW - QA'$, we have

$$x' = OW - QA' = x \cos \theta - y \sin \theta. \tag{10.19}$$

Similarly, we have

$$y' = SP' = SQ + QP' = WA' + QP' = x \sin \theta + y \cos \theta. \tag{10.20}$$

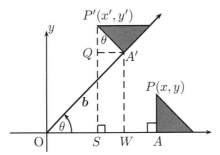

Figure 10.1: The rotational transformation.

The above two equations can be written in a compact form

$$\begin{pmatrix} x' \\ y' \end{pmatrix} = \begin{pmatrix} \cos\theta & -\sin\theta \\ \sin\theta & \cos\theta \end{pmatrix} \begin{pmatrix} x \\ y \end{pmatrix}, \tag{10.21}$$

where the matrix for the rotation or transformation is

$$\boldsymbol{R}_\theta = \begin{pmatrix} \cos\theta & -\sin\theta \\ \sin\theta & \cos\theta \end{pmatrix}. \tag{10.22}$$

Therefore, for any point (x, y), its new coordinates after rotating by an angle θ can be obtained by

$$\begin{pmatrix} x' \\ y' \end{pmatrix} = \boldsymbol{R}_\theta \begin{pmatrix} x \\ y \end{pmatrix}. \tag{10.23}$$

If point P is rotated by $\theta + \psi$, we have

$$\begin{pmatrix} x'' \\ y'' \end{pmatrix} = \begin{pmatrix} \cos(\theta + \psi) & -\sin(\theta + \psi) \\ \sin(\theta + \psi) & \cos(\theta + \psi) \end{pmatrix} = \boldsymbol{R}_{\theta+\psi} \begin{pmatrix} x \\ y \end{pmatrix}, \tag{10.24}$$

which can also be achieved by two steps: first by rotating θ to get $P'(x', y')$ and then rotating by ψ from $P'(x', y')$ to $P''(x'', y'')$. This is to say

$$\begin{pmatrix} x'' \\ y'' \end{pmatrix} = \boldsymbol{R}_\psi \begin{pmatrix} x' \\ y' \end{pmatrix} = \boldsymbol{R}_\psi \boldsymbol{R}_\theta \begin{pmatrix} x \\ y \end{pmatrix}. \tag{10.25}$$

Combining with (10.24), we have

$$\boldsymbol{R}_{\theta+\psi} = \boldsymbol{R}_\psi \boldsymbol{R}_\theta, \tag{10.26}$$

or

$$\begin{pmatrix} \cos(\theta + \psi) & -\sin(\theta + \psi) \\ \sin(\theta + \psi) & \cos(\theta + \psi) \end{pmatrix} = \begin{pmatrix} \cos\psi & -\sin\psi \\ \sin\psi & \cos\psi \end{pmatrix} \begin{pmatrix} \cos\theta & -\sin\theta \\ \sin\theta & \cos\theta \end{pmatrix}$$

$$= \begin{pmatrix} \cos\psi\cos\theta - \sin\psi\sin\theta & -[\cos\psi\sin\theta + \sin\psi\cos\theta] \\ \sin\psi\cos\theta + \cos\psi\sin\theta & \cos\psi\cos\theta - \sin\psi\sin\theta \end{pmatrix}, \quad (10.27)$$

which is another way of deriving the sine and cosine of the addition of two angles.

In a special case when first rotating by θ, followed by rotating back by $-\theta$, a point $P(x, y)$ should reach its original point. That is

$$\begin{pmatrix} x \\ y \end{pmatrix} = \begin{pmatrix} 1 & 0 \\ 0 & 1 \end{pmatrix}\begin{pmatrix} x \\ y \end{pmatrix} = \boldsymbol{R}_{-\theta}\boldsymbol{R}_{\theta}\begin{pmatrix} x \\ y \end{pmatrix}, \quad (10.28)$$

which means that

$$\boldsymbol{R}_{-\theta}\boldsymbol{R}_{\theta} = \begin{pmatrix} 1 & 0 \\ 0 & 1 \end{pmatrix} = \boldsymbol{I}. \quad (10.29)$$

In other words, $\boldsymbol{R}_{-\theta}$ is the inverse of $\boldsymbol{R}_{\theta}$. That is to say

$$\boldsymbol{R}_{-\theta} = \begin{pmatrix} \cos(-\theta) & -\sin(-\theta) \\ \sin(-\theta) & \cos(-\theta) \end{pmatrix} = \begin{pmatrix} \cos\theta & \sin\theta \\ -\sin\theta & \cos\theta \end{pmatrix}, \quad (10.30)$$

is the inverse of

$$\boldsymbol{R}_{\theta} = \begin{pmatrix} \cos\theta & -\sin\theta \\ \sin\theta & \cos\theta \end{pmatrix}. \quad (10.31)$$

In general, the inverse $\boldsymbol{A}^{-1}$ of a square matrix $\boldsymbol{A}$, if it exists, is defined by

$$\boldsymbol{A}^{-1}\boldsymbol{A} = \boldsymbol{A}\boldsymbol{A}^{-1} = \boldsymbol{I}, \quad (10.32)$$

where $\boldsymbol{I}$ is a unit matrix which is the same size as $\boldsymbol{A}$.

Example 10.1:   For example, a 2×2 matrix $\boldsymbol{A}$ and its inverse $\boldsymbol{A}^{-1}$

$$\boldsymbol{A} = \begin{pmatrix} a & b \\ c & d \end{pmatrix}, \qquad \boldsymbol{A}^{-1} = \begin{pmatrix} \alpha & \beta \\ \gamma & \kappa \end{pmatrix},$$

can be related by

$$\boldsymbol{A}\boldsymbol{A}^{-1} = \begin{pmatrix} a & b \\ c & d \end{pmatrix}\begin{pmatrix} \alpha & \beta \\ \gamma & \kappa \end{pmatrix} = \begin{pmatrix} a\alpha + b\gamma & a\beta + b\kappa \\ c\alpha + d\gamma & c\beta + d\kappa \end{pmatrix} = \begin{pmatrix} 1 & 0 \\ 0 & 1 \end{pmatrix} = \boldsymbol{I}.$$

This means that

$$a\alpha + b\gamma = 1, \; a\beta + b\kappa = 0, \; c\alpha + d\gamma = 0, \; c\gamma + d\kappa = 1.$$

These four equations will solve the four unknowns α, β, γ and κ. After some simple rearrangement and calculations, we have

$$\alpha = \frac{d}{\Delta}, \; \beta = \frac{-b}{\Delta}, \; \gamma = -c\Delta, \; \kappa = \frac{a}{\Delta},$$

where $\Delta = ad - bc$ is the determinant of A. Therefore, the inverse becomes

$$A^{-1} = \frac{1}{ad - bc} \begin{pmatrix} d & -b \\ -c & a \end{pmatrix}.$$

It is straightforward to verify that $A^{-1}A = \begin{pmatrix} 1 & 0 \\ 0 & 1 \end{pmatrix}$.

In the case of

$$R_\theta = \begin{pmatrix} \cos\theta & -\sin\theta \\ \sin\theta & \cos\theta \end{pmatrix},$$

we have

$$R^{-1} = \frac{1}{\cos\theta\cos\theta - (-\sin\theta)\sin\theta} \begin{pmatrix} \cos\theta & \sin\theta \\ -\sin\theta & \cos\theta \end{pmatrix} = \begin{pmatrix} \cos\theta & \sin\theta \\ -\sin\theta & \cos\theta \end{pmatrix},$$

where we have used $\cos^2\theta + \sin^2\theta = 1$. This is the same as (10.30).

- -

We have seen that some special combinations of the elements such as the determinant $\Delta = ad - bc$ is very important. We now try to define it more generally.

The determinant of an $n \times n$ square matrix $A = [a_{ij}]$ is a number which can be obtained by cofactor expansion either by row or by column

$$\det(A) \equiv |A| = \sum_{j=1}^{n} (-1)^{i+j} a_{ij} |M_{ij}|, \tag{10.33}$$

where $|M_{ij}|$ is the cofactor or the determinant of a minor matrix M of A, obtained by deleting row i and column j. This is a recursive relationship. For example, M_{12} of a 3×3 matrix is obtained by deleting the first row and the second column

$$\begin{vmatrix} a_{11} - & -a_{12} & a_{13} - \\ a_{21} & a_{22} & a_{23} \\ a_{31} & a_{32} & a_{33} \end{vmatrix} \implies |M|_{12} = \begin{vmatrix} a_{21} & a_{23} \\ a_{31} & a_{33} \end{vmatrix}. \tag{10.34}$$

Obviously, the determinant of a 1×1 matrix $|a_{11}| = a_{11}$ is the number itself. The determinant of a 2×2 matrix

$$\det(A) = \begin{vmatrix} a_{11} & a_{12} \\ a_{21} & a_{22} \end{vmatrix} = a_{11}a_{22} - a_{12}a_{21}. \tag{10.35}$$

The determinant of a 3×3 matrix is given by $\det(A)$ or

$$\begin{vmatrix} a_{11} & a_{12} & a_{23} \\ a_{21} & a_{22} & a_{23} \\ a_{31} & a_{32} & a_{33} \end{vmatrix} = (-1)^{1+1}a_{11} \begin{vmatrix} a_{22} & a_{23} \\ a_{32} & a_{33} \end{vmatrix} + (-1)^{1+2}a_{12} \begin{vmatrix} a_{21} & a_{23} \\ a_{31} & a_{33} \end{vmatrix}$$

$$+ (-1)^{1+3}a_{13} \begin{vmatrix} a_{21} & a_{22} \\ a_{31} & a_{32} \end{vmatrix} = a_{11}(a_{22}a_{33} - a_{32}a_{23})$$

$$-a_{12}(a_{21}a_{33} - a_{31}a_{23}) + a_{13}(a_{21}a_{32} - a_{31}a_{22}). \qquad (10.36)$$

Here we used the expansion along the first row $i = 1$. We can also expand it along any other rows or columns, and the results are the same. As the determinant of a matrix is a scalar or a simple number, it is not difficult to understand the following properties

$$\det(\boldsymbol{AB}) = \det(\boldsymbol{A})\det(\boldsymbol{B}), \qquad \det(\boldsymbol{A}^T) = \det(\boldsymbol{A}). \qquad (10.37)$$

There are many applications of the determinant. For example, $\det(\boldsymbol{A}) = 0$, the square matrix is called singular, and the inverse of such a matrix does not exist. The inverse of a matrix exists only if $\det(\boldsymbol{A}) \neq 0$. Here we will use it to calculate the inverse $\boldsymbol{A}^{-1}$ using

$$\boldsymbol{A}^{-1} = \frac{\mathrm{adj}(\boldsymbol{A})}{\det(\boldsymbol{A})} = \frac{1}{\det(\boldsymbol{A})}\boldsymbol{B}^T, \qquad \boldsymbol{B} = \left[(-1)^{i+j}|M_{ij}|\right], \qquad (10.38)$$

where the matrix $\boldsymbol{B}^T$ is called the adjoint of matrix $\boldsymbol{A}$ with the same size as $\boldsymbol{A}$, and $i, j = 1, ..., n$. Each of the element $\boldsymbol{B}$ is expressed in terms of a cofactor so that $b_{ij} = (-1)^{i+j}|M_{ij}|$. $\boldsymbol{B}$ itself is called the cofactor matrix, while $\mathrm{adj}(\boldsymbol{A}){=}\boldsymbol{B}^T$ is sometimes used to denote the adjoint matrix. This seems too complicated, and let us compute the the inverse of a 3×3 matrix as an example.

Example 10.2: In order to compute the inverse of

$$\boldsymbol{A} = \begin{pmatrix} 1 & 1 & -2 \\ 1 & 0 & 2 \\ 2 & 1 & 1 \end{pmatrix},$$

we first construct its adjoint matrix $\boldsymbol{B}^T$ with

$$\boldsymbol{B} = [b_{ij}] = \left[(-1)^{i+j}|M_{ij}|\right].$$

The first element b_{11} can be obtained by

$$b_{11} = (-1)^{1+1} \begin{vmatrix} 0 & 2 \\ 1 & 1 \end{vmatrix} = (-1)^2 \times (0 \times 1 - 2 \times 1) = -2.$$

The element b_{12} is

$$b_{12} = (-1)^{1+2} \begin{vmatrix} 1 & 2 \\ 2 & 1 \end{vmatrix} = -1 \times (1 \times 1 - 2 \times 2) = 3,$$

while the element b_{21} is

$$b_{21} = (-1)^{2+1} \begin{vmatrix} 1 & -2 \\ 1 & 1 \end{vmatrix} = (-1)^3 \times (1 \times 1 - 1 \times (-2)) = -3.$$

Following a similar procedure, we have B and its transpose B^T as

$$B = \begin{pmatrix} -2 & 3 & 1 \\ -3 & 5 & 1 \\ 2 & -4 & -1 \end{pmatrix}, \quad \text{or} \quad B^T = \begin{pmatrix} -2 & -3 & 2 \\ 3 & 5 & -4 \\ 1 & 1 & -1 \end{pmatrix}.$$

Then, the determinant of A is

$$\det(A) = \begin{vmatrix} 1 & 1 & -2 \\ 1 & 0 & 2 \\ 2 & 1 & 1 \end{vmatrix} = 1 \times \begin{vmatrix} 0 & 2 \\ 1 & 1 \end{vmatrix} - 1 \times \begin{vmatrix} 1 & 2 \\ 2 & 1 \end{vmatrix} + (-2) \times \begin{vmatrix} 1 & 0 \\ 2 & 1 \end{vmatrix}$$

$$= 1 \times (0 \times 1 - 2 \times 1) - 1 \times (1 \times 1 - 2 \times 2) - 2 \times (1 \times 1 - 2 \times 0)$$

$$= 1 \times (-2) - 1 \times (-3) - 2 \times 1 = -1.$$

Finally, the inverse becomes

$$A^{-1} = \frac{B^T}{\det(A)} = \frac{1}{-1} \begin{pmatrix} -2 & -3 & 2 \\ 3 & 5 & -4 \\ 1 & 1 & -1 \end{pmatrix} = \begin{pmatrix} 2 & 3 & -2 \\ -3 & -5 & 4 \\ -1 & -1 & 1 \end{pmatrix}.$$

This result will be used in the next example.

- -

A linear system can be written as a large matrix equation, and the solution of such a linear system will become straightforward if the inverse of a square matrix is used. Let us demonstrate this by an example. For a linear system consisting of three simultaneous equations, we have

$$a_{11}x + a_{12}y + a_{13}z = b_1, \quad a_{21}x + a_{22}y + a_{23}z = b_2, \quad a_{31}x + a_{32}y + a_{33}z = b_3,$$

which can be written as

$$\begin{pmatrix} a_{11} & a_{12} & a_{13} \\ a_{21} & a_{22} & a_{23} \\ a_{31} & a_{32} & a_{33} \end{pmatrix} \begin{pmatrix} x \\ y \\ z \end{pmatrix} = \begin{pmatrix} b_1 \\ b_2 \\ b_3 \end{pmatrix}, \tag{10.39}$$

or more compactly as

$$Au = b, \tag{10.40}$$

where $u = \begin{pmatrix} x & y & z \end{pmatrix}^T$. By multiplying A^{-1} on both sides, we have

$$A^{-1}Au = A^{-1}b. \tag{10.41}$$

Therefore, its solution can be written as $u = A^{-1}b$.

Example 10.3: In order to solve the following system

$$x + y - 2z = -6, \qquad x + 2z = 8, \qquad 2x + y + z = 5,$$

we first write it as $Au = b$, or

$$\begin{pmatrix} 1 & 1 & -2 \\ 1 & 0 & 2 \\ 2 & 1 & 1 \end{pmatrix} \begin{pmatrix} x \\ y \\ z \end{pmatrix} = \begin{pmatrix} -6 \\ 8 \\ 5 \end{pmatrix}.$$

We know from the earlier example that the inverse of A^{-1} is

$$A^{-1} = \begin{pmatrix} 2 & 3 & -2 \\ -3 & -5 & 4 \\ -1 & -1 & 1 \end{pmatrix},$$

we now have $u = A^{-1}b$ or

$$\begin{pmatrix} x \\ y \\ z \end{pmatrix} = \begin{pmatrix} 2 & 3 & -2 \\ -3 & -5 & 4 \\ -1 & -1 & 1 \end{pmatrix} \begin{pmatrix} -6 \\ 8 \\ 5 \end{pmatrix} = \begin{pmatrix} 2 \times (-6) + 3 \times 8 + (-2) \times 5 \\ -3 \times (-6) + (-5) \times 8 + 4 \times 5 \\ -1 \times (-6) + (-1) \times 8 + 1 \times 5 \end{pmatrix} = \begin{pmatrix} 2 \\ -2 \\ 3 \end{pmatrix},$$

which gives a unique set of solutions $x = 2, y = -2$ and $z = 3$.

In general, a linear system of m equations for n unknowns can be written in the compact form as

$$\begin{pmatrix} a_{11} & a_{12} & \cdots & a_{1n} \\ a_{21} & a_{22} & \cdots & a_{2n} \\ \vdots & \vdots & & \\ a_{n1} & a_{n2} & \cdots & a_{nn} \end{pmatrix} \begin{pmatrix} u_1 \\ u_2 \\ \vdots \\ u_n \end{pmatrix} = \begin{pmatrix} b_1 \\ b_2 \\ \vdots \\ b_n \end{pmatrix}, \tag{10.42}$$

or simply

$$Au = b. \tag{10.43}$$

Its solution can be obtained by inverse

$$u = A^{-1}b. \tag{10.44}$$

You may wonder how you can get the inverse A^{-1} of a larger matrix A more efficiently. For large systems, direct inverse is not a good option. There are many other more efficient methods to obtain the solutions, including the powerful Gauss-Jordan elimination, matrix decomposition, and iteration methods. Interested readers can refer to more advanced literature for details.

10.3 Eigenvalues and Eigenvectors

A special case of a linear system $Au = b$ is when $b = \lambda u$, and this becomes an eigenvalue problem. An eigenvalue λ and corresponding eigenvector u of a square matrix A satisfy

$$Au = \lambda u, \quad \text{or}, \quad (A - \lambda I)u = 0. \tag{10.45}$$

Any nontrivial solution requires that

$$\det |\boldsymbol{A} - \lambda \boldsymbol{I}| = 0, \tag{10.46}$$

or

$$\begin{vmatrix} a_{11} - \lambda & a_{12} & \dots & a_{1n} \\ a_{21} & a_{22} - \lambda & \dots & a_{2n} \\ \vdots & \vdots & & \\ a_{n1} & a_{n2} & \dots & a_{nn} - \lambda \end{vmatrix} = 0, \tag{10.47}$$

which is equivalent to

$$\lambda^n + \alpha_{n-1}\lambda^{n-1} + \dots + \alpha_0 = (\lambda - \lambda_1)(\lambda - \lambda_2)\dots(\lambda - \lambda_n) = 0. \tag{10.48}$$

In general, the characteristic equation has n solutions. Eigenvalues have interesting connections with the matrix. The trace of any square matrix is defined as the sum of its diagonal elements, i.e.,

$$\mathrm{tr}(\boldsymbol{A}) = \sum_{i=1}^{n} a_{ii} = a_{11} + a_{22} + \dots + a_{nn}. \tag{10.49}$$

The sum of all the eigenvalues of a square matrix $\boldsymbol{A}$ is equivalent to the trace of $\boldsymbol{A}$. That is

$$\mathrm{tr}(\boldsymbol{A}) = a_{11} + a_{22} + \dots + a_{nn} = \sum_{i=1}^{n} \lambda_i = \lambda_1 + \lambda_2 + \dots + \lambda_n. \tag{10.50}$$

In addition, the eigenvalues are also related to the determinant by

$$\det(\boldsymbol{A}) = \prod_{i=1}^{n} \lambda_i. \tag{10.51}$$

Example 10.4: For a simple 2×2 matrix

$$\boldsymbol{A} = \begin{pmatrix} 1 & 5 \\ 2 & 4 \end{pmatrix},$$

its eigenvalues can be determined by

$$\begin{vmatrix} 1 - \lambda & 5 \\ 2 & 4 - \lambda \end{vmatrix} = 0,$$

or

$$(1 - \lambda)(4 - \lambda) - 2 \times 5 = 0,$$

which is equivalent to

$$(\lambda + 1)(\lambda - 6) = 0.$$

Thus, the eigenvalues are $\lambda_1 = -1$ and $\lambda_2 = 6$. The trace of A is $\text{tr}(A) = a_{11} + a_{22} = 1 + 4 = 5 = \lambda_1 + \lambda_2$. In order to obtain the eigenvector for each eigenvalue, we assume

$$v = \begin{pmatrix} v_1 \\ v_2 \end{pmatrix}.$$

For the eigenvalue $\lambda_1 = -1$, we plug this into

$$|A - \lambda I| v = 0,$$

and we have

$$\begin{vmatrix} 1 - (-1) & 5 \\ 2 & 4 - (-1) \end{vmatrix} \begin{pmatrix} v_1 \\ v_2 \end{pmatrix} = 0, \qquad \begin{vmatrix} 2 & 5 \\ 2 & 5 \end{vmatrix} \begin{pmatrix} v_1 \\ v_2 \end{pmatrix} = 0,$$

which is equivalent to

$$2v_1 + 5v_2 = 0, \qquad \text{or} \qquad v_1 = -\frac{5}{2}v_2.$$

This equation has infinite solutions; each corresponds to the vector parallel to the unit eigenvector. As the eigenvector should be normalised so that its modulus is unity, this additional condition requires

$$v_1^2 + v_2^2 = 1,$$

which means

$$(\frac{-5v_2}{2})^2 + v_2^2 = 1.$$

We have $v_1 = -5/\sqrt{29}$, $v_2 = 2/\sqrt{29}$. Thus, we have the first set of eigenvalue and eigenvector

$$\lambda_1 = -1, \qquad v_1 = \begin{pmatrix} -\frac{5}{\sqrt{29}} \\ \frac{2}{\sqrt{29}} \end{pmatrix}. \tag{10.52}$$

Similarly, the second eigenvalue $\lambda_2 = 6$ gives

$$\begin{vmatrix} 1 - 6 & 5 \\ 2 & 4 - 6 \end{vmatrix} \begin{pmatrix} v_1 \\ v_2 \end{pmatrix} = 0.$$

Using the normalisation condition $v_1^2 + v_2^2 = 1$, the above equation has the following solution

$$\lambda_2 = 6, \qquad v_2 = \begin{pmatrix} \frac{\sqrt{2}}{2} \\ \frac{\sqrt{2}}{2} \end{pmatrix}.$$

Furthermore, the trace and determinant of A are $\text{tr}(A) = 1 + 4 = 5$, and $\det(A) = 1 \times 4 - 2 \times 5 = -6$. The sum of the eigenvalues is $\sum_{i=1}^{2} \lambda_i = -1 + 6 = 5 = \text{tr}(A)$, while the product of the eigenvalues is $\prod_{i=1}^{2} \lambda_i = -1 \times 6 = -6 = \det(A)$. Indeed, the above relationships about eigenvalues are true.

10.4 Harmonic Motion

Harmonic oscillations or mechanical vibrations occur in many processes related to earth sciences. For example, the Earth itself has free oscillations. The natural frequencies of a system can often be calculated using eigenvalue methods because the natural frequencies are the eigenvalues if the system equations are formulated properly. Let us demonstrate this using a simple system with three mass blocks connected by two springs as shown in Figure 10.2. This system can be thought of as a tectonic plate pushing two other plates, or a car towing two caravans on a flat road, ignoring friction.

Let u_1, u_2, u_3 be the displacement of the three mass blocks m_1, m_2, m_3, respectively. Then, their accelerations will be $\ddot{u}_1$, $\ddot{u}_2$, $\ddot{u}_3$ where $\ddot{u} = d^2u/dt^2$. From the balance of forces and Newton's law, we have

$$m_1\ddot{u}_1 = k_1(u_2 - u_1), \tag{10.53}$$

$$m_2\ddot{u}_2 = k_2(u_3 - u_2) - k_1(u_2 - u_1), \tag{10.54}$$

$$m_3\ddot{u}_3 = -k_2(u_3 - u_2). \tag{10.55}$$

These equations can be written in a matrix form as

$$\begin{pmatrix} m_1 & 0 & 0 \\ 0 & m_2 & 0 \\ 0 & 0 & m_3 \end{pmatrix} \begin{pmatrix} \ddot{u}_1 \\ \ddot{u}_2 \\ \ddot{u}_3 \end{pmatrix} + \begin{pmatrix} k_1 & -k_1 & 0 \\ -k_1 & k_1 + k_2 & -k_2 \\ 0 & -k_2 & k_2 \end{pmatrix} \begin{pmatrix} u_1 \\ u_2 \\ u_3 \end{pmatrix} = \begin{pmatrix} 0 \\ 0 \\ 0 \end{pmatrix},$$

or

$$\boldsymbol{M}\ddot{\boldsymbol{u}} + \boldsymbol{K}\boldsymbol{u} = 0, \tag{10.56}$$

where $\boldsymbol{u}^T = (u_1, u_2, u_3)$. The mass matrix $\boldsymbol{M}$ and stiffness matrix $\boldsymbol{K}$ are

$$\boldsymbol{M} = \begin{pmatrix} m_1 & 0 & 0 \\ 0 & m_2 & 0 \\ 0 & 0 & m_3 \end{pmatrix}, \qquad \boldsymbol{K} = \begin{pmatrix} k_1 & -k_1 & 0 \\ -k_1 & k_1 + k_2 & -k_2 \\ 0 & -k_2 & k_2 \end{pmatrix}. \tag{10.57}$$

Equation (10.56) is a second-order ordinary differential equation in terms of matrices. We will learn more about ordinary differential equations in the next chapter. At the moment, we assume that the motion of our system is harmonic, therefore, we write their solution in the form $\boldsymbol{u} = \boldsymbol{U}\cos(\omega t)$ where $\boldsymbol{U} = (U_1, U_2, U_3)^T$ is a constant vector related to the amplitudes of the vibrations. Here the unknown ω is the natural frequency or frequencies of the system.

We know that $\ddot{\boldsymbol{u}} = -\boldsymbol{U}\omega^2\cos(\omega t)$. Then, equation (10.56) becomes

$$(\boldsymbol{K} - \omega^2\boldsymbol{M})\boldsymbol{U} = 0. \tag{10.58}$$

This is essentially an eigenvalue problem because any non-trivial solutions for $\boldsymbol{U}$ require

$$|\boldsymbol{K} - \omega^2\boldsymbol{M}| = 0. \tag{10.59}$$

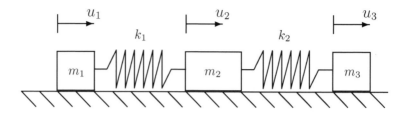

Figure 10.2: Harmonic vibrations and their natural frequencies.

Therefore, the eigenvalues of this equation give the natural frequencies.

Example 10.5: For the simplest case when $m_1 = m_2 = m_3 = m$ and $k_1 = k_2 = k$, we have

$$\begin{vmatrix} k - \omega^2 m & -k & 0 \\ -k & 2k - \omega^2 m & -k \\ 0 & -k & k - \omega^2 m \end{vmatrix} = 0,$$

or

$$-\omega^2 (k - \omega^2 m)(3km - \omega^2 m^2) = 0.$$

This is a cubic equation in terms of ω^2, and it has three solutions. Therefore, the three natural frequencies are

$$\omega_1^2 = 0, \qquad \omega_2^2 = \frac{k}{m}, \qquad \omega_3^2 = \frac{3k}{m}.$$

For $\omega_1^2 = 0$, we have $U^T = (U_1, U_2, U_3) = \frac{1}{\sqrt{3}}(1, 1, 1)$, which is the rigid body motion. For $\omega_2 = k/m$, the eigenvector is determined by

$$\begin{pmatrix} 0 & -k & 0 \\ -k & k & -k \\ 0 & -k & 0 \end{pmatrix} \begin{pmatrix} U_1 \\ U_2 \\ U_3 \end{pmatrix} = \begin{pmatrix} 0 \\ 0 \\ 0 \end{pmatrix},$$

which leads to $U_2 = 0$, and $U_1 = U_3$. Written in normalised form, it becomes $(U_1, U_2, U_3) = \frac{1}{\sqrt{2}}(1, 0, -1)$. This means that block 1 moves in the opposite direction away from block 3, and block 2 remains stationary. For $\omega_3^2 = 3k/m$, we have $(U_1, U_2, U_3) = \frac{1}{\sqrt{6}}(1, -2, 1)$. That is to say, block 2 moves in the different direction from block 3 which is at the same pace with block 1.

Chapter 11

Ordinary Differential Equations

Differential equations are very important in science, and many geophysical and geological processes can be modelled in terms of differential equations. In this chapter, we will introduce ordinary differential equations and their basic solution techniques, and in the next chapter will discuss the more complicated partial differential equations.

Differential equations have been applied to almost every branch of earth sciences. As illustrative examples for this chapter, we will apply them to study the variations of air pressure with altitude, climate changes, flexural deflection of the lithosphere, and post-glacier isostatic adjustment.

11.1   Differential Equations

In the introduction to basic equations such as $x^3 - x^2 + x - 1 = 0$, we know that the relationship is a function $f(x) = x^3 - x^2 + x - 1$ and the only unknown is x. The aim is to find values of x which satisfy $f(x) = 0$. It is easy to verify that the equation has three solutions $x = 1, \pm i$.

A differential equation, on the other hand, is a relationship that contains functions and their derivatives. For example, the following equation

$$\frac{dy}{dx} = x^3 - x, \tag{11.1}$$

is a differential equation because it provides a relationship between the derivative dy/dx and the function $f(x) = x^3 - x$. The unknown is a function $y(x)$ and the aim is to find a function $y(x)$ (not a simple value) which satisfies the above equation. Here x is the independent variable.

149

From the integration, we know that the gradient dy/dx is $x^3 - x$, and the function $\frac{1}{4}x^4 - \frac{1}{2}x^2$ has the gradient $x^3 - x$. We can say that $y(x) = \frac{x^4}{4} - \frac{x^2}{2}$ satisfies the differential equation (11.1), and thus $y(x)$ is a solution to (11.1). Then the question is: Are there any other solutions to this equation? The answer is an infinite number. From Fig. 6.2, we know that there is a family of curves whose gradient is $x^3 - x$. The solution in general should contain an arbitrary constant C. That is to say, the general solution of (11.1) can be written as

$$y(x) = \frac{x^4}{4} - \frac{x^2}{2} + C. \tag{11.2}$$

Any solution that corresponds to a single specific curve is a particular solution. For example, both $x^4/4 - x^2/2$ and $x^4/4 - x^2/2 - 1$ are particular solutions.

Another important concept is the order of a differential equation. The order of a differential equation is the highest derivative of the unknown. For example, the order of (11.1) is 1 as the highest derivative is the gradient, so this equation is called a first-order differential equation. The following equation

$$\frac{d^2y(x)}{dx^2} - 2x\frac{dy(x)}{dx} + y(x) = x^2, \tag{11.3}$$

is the second-order differential equation as the highest derivative is the second derivative d^2y/dx^2.

All the above equations only contain first and/or second derivatives, and there is only a single independent variable x. Such differential equations are called ordinary differential equations (ODE).

On the other hand, if a quantity of interest such as u depends on two or more independent variables such x and t, then partial derivatives such as $\frac{\partial u}{\partial x}$ and $\frac{\partial u}{\partial t}$ will appear. In this case, we usually have to deal with a partial differential equation (PDE). For example, the following equation

$$\frac{\partial u}{\partial t} = \frac{\partial^2 u}{\partial x^2}, \tag{11.4}$$

is a diffusion equation and we will discuss this in more detail in the next chapter. In this chapter, we will focus solely on the ordinary differential equations.

11.2   First-Order Equations

We know that $dy/dx = x^3 - x$ is a first-order ordinary differential equation, which can be generalised as

$$\frac{dy(x)}{dx} = f(x), \tag{11.5}$$

where $f(x)$ is a given function of x. Its solution can be obtained by simple integration

$$y(x) = \int f(x)dx + C. \tag{11.6}$$

Example 11.1: The air can be considered as an ideal gas obeying the ideal gas law

$$pV = nRT,$$

where P is the air pressure, V is the volume of the air of interest, and n is the number of modes of air. $R = 8.31$ J/mole·K is the universal gas constant, and T is the absolute temperature. The ideal gas law can also be written as

$$p = \frac{\rho R}{M} T,$$

where ρ is the density of the air and $M \approx 28.9$ g/mole is the molar mass of the air.

We assume that the air pressure is hydrostatic. That is, the increase of the pressure dp is balanced by the increment of the weight $-\rho g dz$ for a thin layer with a unit area. Here z is the altitude above the Earth's surface and the $-$ sign indicates the fact that the z increases and pressure decreases in atmosphere. Therefore, we have $dp = -\rho g dz$ or

$$\frac{dp}{dz} = -\rho g.$$

From $p = \rho RT/M$, we have $\rho = pM/RT$, and we now have

$$\frac{dp}{dz} = -\frac{pMg}{RT},$$

or

$$\frac{dp}{p} = -\frac{Mg}{RT}dz.$$

Integrating from $z = 0$ to $z = h$, we have

$$\int_{p_0}^{p} dp = \ln p - \ln p_0 = -\int_{0}^{h} \frac{Mg}{RT}dz = -\frac{Mg}{RT}h,$$

where p_0 is the pressure on the Earth's surface at $z = 0$.

This means that

$$\ln \frac{p}{p_0} = -\frac{Mg}{RT}h.$$

Taking the logarithms, we have

$$p = p_0 e^{-\gamma h}, \qquad \gamma = \frac{Mg}{RT}.$$

We can see that the air pressure decreases exponentially as the height h increases. We can define a characteristic height

$$L = \frac{1}{\beta} = \frac{RT}{Mg},$$

so that

$$p = p_0 e^{-h/L}.$$

For the typical values of $M = 0.0289$ kg/mole, $g = 9.8$ m/s^2, and $T = 293$ K (or 20°C), we have

$$L = \frac{8.31 \times 393}{0.0289 \times 9.8} \approx 8597 \text{ m} = 8.597 \text{ km},$$

which corresponds to $\gamma \approx 0.00116$ m^{-1}. This means that the air pressure at $z = L$ will becomes $1/e \approx 36.8\%$ of the pressure on the Earth's surface.

The exponential decrease of the air pressure was conventionally used to determine the altitude in aviation and atmospheric sciences. From $p = p_0 e^{-\gamma h}$, we get

$$\ln\left(\frac{p}{p_0}\right) = -\gamma h,$$

which leads to

$$h = -\frac{1}{\gamma} \ln \frac{p}{p_0} = L \ln \frac{p_0}{p}.$$

So if the pressure has dropped to half of the ground pressure (assuming a constant temperature), the altitude h is given by

$$h = L \ln \frac{p_0}{0.5 p_0} = L \ln 2 \approx 5.959 \text{ km}.$$

In our calculations, we have assumed that the temperature is constant. In reality, the temperature does change significantly.

What happens if f also depends on y, so that we have

$$\frac{dy(x)}{dx} = f(x, y). \tag{11.7}$$

It is not possible to solve it by integration in general; however, in a special case when $f(x, y) = g(x)h(y)$ is separable we can solve it by the separation of variables. That is to rearrange the equation so that the function of x is on one side, and the function of y is on the other side. We have

$$\frac{1}{h(y)} \frac{dy}{dx} = g(x). \tag{11.8}$$

Then, we can integrate both sides with respect to x, and we have

$$\int \frac{1}{h(y)} \frac{dy}{dx} dx = \int \frac{1}{h(y)} dy = \int g(x) dx, \tag{11.9}$$

where we have used $dy = (dy/dx)dx$ or $\int() \frac{dy}{dx} dx = \int()dy$. If we provide appropriate conditions for y or $y'(0)$ at $x = 0$, we can also determine the unknown constant. Let us look at an example.

Example 11.2: To solve the first-order differential equation

$$\frac{dy}{dx} = (x^4 + e^x)e^{-y},$$

under the condition $y(0) = 0$ at $x = 0$, we can rewrite it as

$$\frac{1}{e^{-y}} \frac{dy}{dx} = x^4 + e^x,$$

whose integration leads to

$$\int \frac{1}{e^{-y}} \frac{dy}{dx} dx = \int (x^4 + e^x) dx.$$

This becomes

$$\int e^y dy = \int x^4 dx + \int e^x dx,$$

or

$$e^y = \frac{x^5}{5} + e^x + C,$$

Using $x = 0$ and $y(0) = 0$, we have

$$e^0 = \frac{0^5}{5} + e^0 + C,$$

which gives $1 = 0 + 1 + C$ or $C = 0$. Now the solution becomes

$$e^y = \frac{x^5}{5} + e^x.$$

Taking the logarithm, we have

$$y = \ln(\frac{x^5}{5} + e^x).$$

We will use this solution in the numerical solution of ordinary differential equations in later chapters.

When the function $f(x, y)$ is a nonlinear function of y such as the case of e^{-y}, it is not so straightforward to solve such equations. In the rest of the section, we will only focus on the general linear first-order ordinary differential equation in the following form

$$\frac{dy}{dx} + p(x)y = q(x), \tag{11.10}$$

where $p(x)$ and $q(x)$ are known functions of x. We will attempt to solve this generic differential equation. How do we start? We know from the differentiation rule for a product that

$$\frac{d(Q(x)y)}{dx} = Q(x)\frac{dy}{dx} + y\frac{dQ(x)}{dx} = Q(x)y' + Q'(x)y. \qquad (11.11)$$

If we could in some way write the left hand side of (11.10) as a derivative of a product similar to the above, we should be able to rewrite the original differential equation (11.10) as a simpler form (11.5). Therefore, multiplying both sides of (11.10) by a function $Q(x)$ to be determined later, we have

$$Q(x)\frac{dy}{dx} + Q(x)p(x)y = Q(x)q(x). \qquad (11.12)$$

The aim is to write it as the form

$$\frac{d(Q(x)y)}{dx} = Q(x)\frac{dy}{dx} + \frac{dQ(x)}{dx}y = Q(x)q(x). \qquad (11.13)$$

By comparing with (11.12), we get

$$\frac{dQ(x)}{dx} = Q(x)p(x). \qquad (11.14)$$

This is a separable form as discussed earlier. We have

$$\frac{1}{Q(x)}\frac{dQ}{dx} = p(x), \qquad (11.15)$$

or

$$\int \frac{1}{Q(x)}\frac{dQ}{dx}dx = \int p(x)dx. \qquad (11.16)$$

This gives

$$\ln Q(x) = \int p(x)dx, \qquad (11.17)$$

or

$$Q(x) = e^{\int p(x)dx}, \qquad (11.18)$$

which is often called the integrating factor. Using this integrating factor, equation (11.12) becomes

$$\frac{d[e^{\int p(x)dx}y]}{dx} = Q(x)q(x) = q(x)e^{\int p(x)dx}. \qquad (11.19)$$

Integrating it with respect to x, we have

$$e^{\int p(x)dx}y = \int \left[q(x)e^{\int p(x)dx}\right]dx + A, \qquad (11.20)$$

where A is the constant of integration. Dividing both sides of the above equation by the integrating factor, we have the general solution

$$y = e^{-\int p(x)dx} \left\{ \int \left[q(x)e^{\int p(x)dx} \right] dx \right\} + Ae^{-\int p(x)dx}. \qquad (11.21)$$

Let us solve the differential equation

$$\frac{dy}{dx} + 5y = e^{-x}, \qquad (11.22)$$

given that $y(0) = 1/2$. The integrating factor $Q(x)$ is

$$Q(x) = e^{\int 5dx} = e^{5x}. \qquad (11.23)$$

Multiplying the original equation (11.22) by $Q(x)$, so we have

$$e^{5x}\frac{dy}{dx} + 5e^{5x}y = \frac{d(e^{5x}y)}{dx} = e^{5x}e^{-x} = e^{4x}, \qquad (11.24)$$

whose integration leads to

$$e^{5x}y = \frac{e^{4x}}{4} + A. \qquad (11.25)$$

Therefore, the general solution is

$$y = \frac{e^{4x}}{4e^{5x}} + \frac{A}{e^{5x}} = \frac{1}{4}e^{-x} + Ae^{-5x}. \qquad (11.26)$$

The constant A can be determined by $y(0) = 1/2$ at $x = 0$. That is

$$\frac{1}{2} = \frac{1}{4}e^{-0} + Ae^{-5\times0}, \qquad (11.27)$$

or $A = 1/4$. The final solution becomes

$$y = \frac{1}{4}e^{-x} + \frac{1}{4}e^{-5x}. \qquad (11.28)$$

Let us look at another example.

Example 11.3: In hydrology, the runoff model for linear reservoir provides a flow equation

$$\frac{dq}{dt} + \alpha q = \alpha R,$$

where q is the runoff or discharge, and R is the effective rainfall and can be considered as a constant for a short time. The coefficient α is a response factor which can be taken as a constant.

We know that the integrating factor is

$$Q = e^{\int \alpha dt} = e^{\alpha t}.$$

Therefore, using the formula (11.21), we have the general solution

$$q = e^{-\alpha t} \int (\alpha R) e^{\alpha t} dt + A e^{-\alpha t} = e^{-\alpha t} (\alpha R) \frac{e^{\alpha t}}{\alpha} + A e^{-\alpha t},$$

which becomes

$$q = A e^{-\alpha t} + R.$$

If the discharge at $t = 0$ is q_0, we have

$$q_0 = A e^{-\alpha \times 0} + R,$$

or

$$A = q_0 - R.$$

The final solution now becomes

$$q = q_0 e^{-\alpha t} + R(1 - e^{-\alpha t}).$$

As $t \to \infty$, we have $q \to R$.

11.3 Second-Order Equations

For second-order ordinary differential equations (ODEs), it is generally more tricky to find the general solution. However, a special case with significantly practical importance and mathematical simplicity is the second-order linear differential equation with constant coefficients in the following form

$$\frac{d^2y}{dx^2} + b\frac{dy}{dx} + cy(x) = f(x), \tag{11.29}$$

where the coefficients b and c are constants, and $f(x)$ is a known smooth function of x. Obviously, the more general form would be

$$a\frac{d^2y}{dx^2} + b\frac{dy}{dx} + cy(x) = f(x), \tag{11.30}$$

however, if we divide both sides by a, we will reach our standard form. Here we assume $a \neq 0$. In a special case of $a = 0$, it reduces to a first-order linear differential equation which has been discussed in the previous section. So we will start our discussion from (11.29).

A differential equation is said to be homogeneous if $f(x) = 0$. For a given generic second-order differential equation (11.29), a function that satisfies the homogeneous equation

$$\frac{d^2y}{dx^2} + b\frac{dy}{dx} + cy(x) = 0, \qquad (11.31)$$

is called the complementary function, denoted by $y_c(x)$. Obviously, the complementary function y_c alone cannot satisfy the original equation (11.29) because there is no way to produce the required $f(x)$ on the right-hand side. Therefore, we have to find a specific function, $y_*(x)$ called the particular integral, so that it indeed satisfies the original equation (11.29). The combined general solution

$$y(x) = y_c(x) + y_*(x), \qquad (11.32)$$

will automatically satisfy the original equation (11.29). The general solution of (11.29) consists of two parts: the complementary function $y_c(x)$ and the particular integral $y_*(x)$. We can obtain these two parts separately, and simply add them together because the original equation is linear, so their solutions are linear combinations.

First things first, how to obtain the complementary function? The general technique is to assume that it takes the form

$$y_c(x) = Ae^{\lambda x}, \qquad (11.33)$$

where A is a constant, and λ is an exponent to be determined. Substituting this assumed form into the homogeneous equation, we have

$$A\lambda^2 e^{\lambda x} + bA\lambda e^{\lambda x} + cAe^{\lambda x} = 0. \qquad (11.34)$$

Since $Ae^{\lambda x}$ should not be zero (otherwise, we have a trivial solution $y_c = 0$ everywhere), we can divide all the terms by $Ae^{\lambda x}$, and we have

$$\lambda^2 + b\lambda + c = 0, \qquad (11.35)$$

which is the characteristic equation for the homogeneous equation. It is also called the auxiliary equation of the ODE. The solution of λ in this case is simply

$$\lambda = \frac{-b \pm \sqrt{b^2 - 4c}}{2}. \qquad (11.36)$$

For simplicity, we can take $A = 1$ as it does not affect the results.

In the discussion of quadratic equations in the first chapter of the book, we know that there are three possibilities for λ. They are: I) two real distinct roots, II) two identical roots, and III) two complex roots.

In the case of two different roots: $\lambda_1 \neq \lambda_2$. Then, both $e^{\lambda_1 x}$ and $e^{\lambda_2 x}$ satisfy the homogeneous equation, so their linear combination forms the complementary function

$$y_c(x) = Ae^{\lambda_1 x} + Be^{\lambda_2 x}, \qquad (11.37)$$

where A and B are constants.

In the special case of identical roots $\lambda_1 = \lambda_2$, or

$$c = \lambda_1^2, \qquad b = -2\lambda_1. \tag{11.38}$$

we cannot simply write

$$y_c(x) = Ae^{\lambda_1 x} + Be^{\lambda_1 x} = (A + B)e^{\lambda_1 x}, \tag{11.39}$$

because it still only one part of the complementary function $y_1 = Ce^{\lambda_1}$ where $C = A + B$ is just another constant. In this case, we should try a different combination, say, $y_2 = xe^{\lambda_1 x}$ to see if it satisfies the homogeneous equation or not. Since $y_2'(x) = e^{\lambda_1 x} + x\lambda_1 e^{\lambda_1 x}$, and $y_2''(x) = \lambda_1 e^{\lambda_1 x} + \lambda_1 e^{\lambda_1 x} + x\lambda_1^2 e^{\lambda_1 x}$, we have

$$y_2''(x) + by_2'(x) + cy_2(x) = e^{\lambda_1 x}(2\lambda_1 + x\lambda_1^2)e^{\lambda_1 x} + be^{\lambda_1 x}(1 + x\lambda_1) + cxe^{\lambda_1 x}$$

$$= e^{\lambda_1 x}[(2\lambda_1 + b)] + xe^{\lambda_1 x}[\lambda_1^2 + b\lambda_1 + c] = 0, \tag{11.40}$$

where we have used $b + 2\lambda_1 = 0$ (identical roots) and $\lambda_1^2 + b\lambda_1 + c = 0$ (the auxiliary equation). This indeed implies that $xe^{\lambda_1 x}$ also satisfies the homogeneous equation. Therefore, the complementary function for the identical roots is

$$y_c(x) = Ae^{\lambda_1 x} + Bxe^{\lambda_1 x} = (A + Bx)e^{\lambda_1 x}. \tag{11.41}$$

Example 11.4: The second-order homogeneous equation

$$\frac{d^2 y}{dx^2} + 5\frac{dy}{dx} - 6y = 0,$$

has a corresponding auxiliary equation

$$\lambda^2 + 5\lambda - 6 = (\lambda - 2)(\lambda + 3) = 0.$$

It has two real roots

$$\lambda_1 = 2, \qquad \lambda_2 = -3.$$

So the complementary function is

$$y_c(x) = Ae^{2x} + Be^{-3x}.$$

But for the differential equation

$$\frac{d^2 y}{dx^2} + 6\frac{dy}{dx} + 9y(x) = 0,$$

its auxiliary equation becomes

$$\lambda^2 + 6\lambda + 9 = 0,$$

which has two identical roots $\lambda_1 = \lambda_2 = -3$. The complementary function in this case can be written as

$$y_c(x) = (A + Bx)e^{-3x}.$$

As complex roots always come in pairs, the case of complex roots would give

$$\lambda_{1,2} = \alpha \pm i\beta, \tag{11.42}$$

where α and β are real numbers. The complementary function becomes

$$y_c(x) = Ae^{(\alpha+i\beta)x} + Be^{(\alpha-i\beta)x} = Ae^{\alpha x}e^{i\beta x} + Be^{\alpha x}e^{-i\beta x}$$

$$= e^{\alpha x}[Ae^{i\beta x} + Be^{-i\beta x}]$$

$$= e^{\alpha x}\{A[\cos(\beta x) + i\sin(\beta x)] + B[\cos(-\beta x) + i\sin(-\beta x)]\}$$

$$= e^{\alpha x}[(A + B)\cos(\beta x) + i(A - B)\sin(\beta x)]$$

$$= e^{\alpha x}[C\cos\beta x + D\sin\beta x], \tag{11.43}$$

where we have used the Euler formula $e^{\theta i} = \cos\theta + i\sin\theta$ and absorb the constants A and B into $C = A + B$ and $D = (A - B)i$.

A special case is when $\alpha = 0$, so the roots are purely imaginary. We have $b = 0$, and $c = \beta^2$. Equation (11.29) in this case becomes

$$\frac{d^2y}{dx^2} + \beta^2 y = 0, \tag{11.44}$$

which is a differential equation for harmonic motions such as the oscillations of a pendulum or a small-amplitude seismic detector. Here β is the angular frequency of the system.

Example 11.5:   For a simple pendulum of mass m shown in Figure 11.1, we now try to derive its equation of oscillations and its period.

Since the motion is circular, the tension or the centripetal force T is thus given by

$$T = m\frac{v^2}{L} = m\dot{\theta}^2 L,$$

where $\dot{\theta} = d\theta/dt$ is the angular velocity.

Forces must be balanced both vertically and horizontally. The component of T in the vertical direction is $T\cos\theta$ which must be equivalent to mg, though in the opposite direction. Here g is the acceleration due to gravity. That is

$$T\cos\theta = mg.$$

Since θ is small or $\theta \ll 1$, we have $\cos\theta \approx 1$. This means that $T \approx mg$.

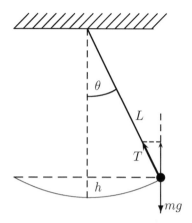

Figure 11.1: A simple pendulum and its harmonic motion.

In the horizontal direction, Newton's second law $F = ma$ implies that the horizontal force $-T \sin \theta$ must be equal to the mass m times the acceleration $L \frac{d^2\theta}{dt^2}$. Now we have

$$m(L\frac{d^2\theta}{dt^2}) = -T \sin \theta \approx -mg \sin \theta.$$

Dividing both sides by mL, we have

$$\frac{d^2\theta}{dt^2} + \frac{g}{L} \sin \theta = 0.$$

Since θ is small, we have $\sin \theta \approx \theta$. Therefore, we finally have

$$\frac{d^2\theta}{dt^2} + \frac{g}{L}\theta = 0.$$

This is the equation of motion for a simple pendulum. From equation (11.44), we know that the angular frequency is $\omega^2 = g/L$ or $\omega = \sqrt{\frac{g}{L}}$. Thus the period of the pendulum is

$$T = \frac{2\pi}{\omega} = 2\pi \sqrt{\frac{L}{g}}. \tag{11.45}$$

We can see that the period is independent of the bob mass. For $L = 1$ m and $g = 9.8$ m/s^2, the period is approximately $T = 2\pi \sqrt{\frac{1}{9.8}} \approx 2$ s.

- -

Now we will try to find the particular integral $y_*(x)$ for the non-homogeneous equation. For particular integrals, we do not intend to find the general form; any specific function or integral that satisfies the original equation (11.29) will do. Before we can determine the particular integral, we have to use some trial functions, and such functions will have strong similarity with the function $f(x)$. For example, if $f(x)$ is a polynomial such as $x^2 + \alpha x + \beta$, we will try a similar form $y_*(x) = ax^2 + bx + c$ and try to determine the coefficients. Let us demonstrate this by an example.

Example 11.6: In order to solve the differential equation

$$\frac{d^2y}{dx^2} + 5\frac{dy}{dx} - 6y = x - 2,$$

we first find its complementary function. From the earlier example, we know that the complementary function can be written as

$$y_c(x) = Ae^{2x} + Be^{-3x}.$$

For the particular integral, we know that $f(x) = x - 2$, so we try the form $y_* = ax + b$. Substituting it into the original equation

$$0 + 5a - 6(ax + b) = x - 2,$$

or

$$(-6a)x + (5a - 6b) = x - 2.$$

As this equality must be true for any x, so the coefficients of the same power of x on both sides of the equation should be equal. That is

$$-6a = 1, \qquad (5a - 6b) = -2,$$

which gives $a = -\frac{1}{6}$, and $b = \frac{7}{36}$. So the general solution becomes

$$y(x) = Ae^{2x} + Be^{-3x} - \frac{x}{6} + \frac{7}{36}.$$

- -

Similarly, if $f(x) = e^{\alpha x}$, we will try to $y_*(x) = ae^{\alpha x}$ so as to determine a. In addition, we $f(x) = \sin \alpha x$ or $\cos \alpha x$, we will attempt the general form $y_*(x) = a \cos \alpha x + b \sin \alpha x$.

11.4   Higher-Order ODEs

Higher-order ODEs are more complicated to solve even when they are linear. In the special case of higher-order ODEs where all the coefficients $a_n, ..., a_1, a_0$ are constants,

$$a_n y^{(n)} + ... + a_1 y' + a_0 y = f(x), \tag{11.46}$$

the solution procedure is identical to that for the second-order differential equation we just discussed in the previous section. The general solution $y(x)$ again consists of two parts: the complementary function $y_c(x)$ and the particular integral or particular solution $y_p^*(x)$. We have

$$y(x) = y_c(x) + y_p^*(x). \tag{11.47}$$

The complementary function which is the solution of the linear homogeneous equation with constant coefficients can be written in a generic form

$$a_n y_c^{(n)} + a_{n-1} y_c^{(n-1)} + ... + a_1 y_c' + a_0 = 0. \tag{11.48}$$

Assuming $y = Ae^{\lambda x}$ where A is a constant, we get the characteristic equation as a polynomial

$$a_n \lambda^n + a_{n-1} \lambda^{(n-1)} + ... + a_1 \lambda + a_0 = 0, \tag{11.49}$$

which has n roots in the general case. Then, the solution can be expressed as the summation of various terms $y_c(x) = \sum_{k=1}^{n} c_k e^{\lambda_k x}$ if the polynomial has n distinct zeros $\lambda_1, ...\lambda_n$. For complex roots, and complex roots always occur in pairs $\lambda = r \pm i\omega$, the corresponding linearly independent terms can then be replaced by $e^{rx}[A\cos(\omega x) + B\sin(\omega x)]$.

The particular solution $y_p^*(x)$ is any $y(x)$ that satisfies the original inhomogeneous equation (11.46). Depending on the form of the function $f(x)$, the particular solutions can take various forms. For most of the combinations of basic functions such as $\sin x$, $\cos x$, e^{kx}, and x^n, the method of undetermined coefficients is widely used. For $f(x) = \sin(\alpha x)$ or $\cos(\alpha x)$, then we can try $y_p^* = A\sin\alpha x + B\sin\alpha x$. We then substitute it into the original equation (11.46) so that the coefficients A and B can be determined. For a polynomial $f(x) = x^n (n = 0, 1, 2, ..., N)$, we then try $y_p^* = A + Bx + ... + Qx^n$ (polynomial). For $f(x) = e^{kx}x^n$, $y_p^* = (A + Bx + ...Qx^n)e^{kx}$. Similarly, $f(x) = e^{kx}\sin\alpha x$ or $f(x) = e^{kx}\cos\alpha x$, we can use $y_p^* = e^{kx}(A\sin\alpha x + B\cos\alpha x)$. More general cases and their particular solutions can be found in various textbooks.

A very useful technique is to use the method of differential operator D. A differential operator D is defined as

$$D \equiv \frac{d}{dx}. \tag{11.50}$$

Since we know that $De^{\lambda x} = \lambda e^{\lambda x}$ and $D^n e^{\lambda x} = \lambda^n e^{\lambda x}$, so they are equivalent to $D \mapsto \lambda$, and $D^n \mapsto \lambda^n$. Thus, any polynomial $P(D)$ will map to $P(\lambda)$. On the other hand, the integral operator $D^{-1} = \int dx$ is just the inverse of the differentiation. The beauty of the differential operator form is that it can be factorised in the same way as for a polynomial, then it can solved using each factor separately. The differential operator is very useful in finding out both the complementary functions and particular integral.

Example 11.7: To find the particular integral for the equation

$$y''''''' + 22y = 15e^{2x}, \tag{11.51}$$

we get

$$(D^7 + 22)y_p^* = 15e^{2x}, \tag{11.52}$$

or

$$y_p^* = \frac{15}{(D^7 + 22)}e^{2x}. \tag{11.53}$$

Since $D^7(e^{2x}) = 2^7 e^{2x}$ or $D^7 \mapsto \lambda^7 = 2^7$, we have

$$y_p^* = \frac{15e^{2x}}{2^7 + 22} = \frac{15e^{2x}}{150} = \frac{e^{2x}}{10}. \tag{11.54}$$

This method also works for $\sin x, \cos x, \sinh x$ and others, and this is because they are related to $e^{\lambda x}$ via $\sin\theta = \frac{1}{2i}(e^{i\theta} - e^{-i\theta})$ and $\cosh x = (e^x + e^{-x})/2$.

Higher-order differential equations can conveniently be written as a system of differential equations. In fact, an nth-order linear equation can always be written as a linear system of n first-order differential equations. A linear system of ODEs is more suitable for mathematical analysis and numerical integration.

11.5 Applications

We have studied the solution of ordinary differential equations. Let us see how they are applied in modelling real-world problems.

11.5.1 Climate Changes

A simple CO_2 model with a constant rate w of emission is governed by

$$\frac{d^2 A}{dt^2} + p\frac{dA}{dt} + qA = rw, \tag{11.55}$$

where $A(t) = C - A_0$ is the excess CO_2 concentration in the atmosphere. That is the difference between the actual CO_2 concentration C above the pre-industrial CO_2 concentration A_0 with an average value of $A_0 = 280$ ppm. p and q are the combined reservoir transfer coefficents, and r is a constant. This simple model is based on the models by Holter *et al* and Keeling's three-reservoir model for carbon dioxide cycle.

By assuming that $A = e^{\lambda t}$ and substituting into the homogeneous equation

$$\frac{d^2 A}{dt^2} + p\frac{dA}{dt} + qA = 0, \tag{11.56}$$

we have

$$\lambda^2 + p\lambda + q = 0, \tag{11.57}$$

whose solution is

$$\lambda_1 = -\frac{p}{2}\left(1 + \sqrt{1 - \frac{4q}{p^2}}\right), \qquad \lambda_2 = -\frac{p}{2}\left(1 - \sqrt{1 - \frac{4q}{p^2}}\right). \tag{11.58}$$

For typical values of $p = 1.007/\text{year}$, $q = 0.0123/\text{year}^2$, we have

$$\lambda_1 \approx -0.995, \qquad \lambda_2 \approx -0.0124. \tag{11.59}$$

Therefore, the homogeneous solution can be written as

$$A = ae^{\lambda_1 t} + be^{\lambda_2 t}. \tag{11.60}$$

Since $|\lambda_1| \gg |\lambda_2|$, the first term with $e^{-\lambda_1 t}$ will become very small if t is large. In climate dynamics, we are often dealing with the timescale of hundreds of years, so we can essentially use the approximate solution

$$A \approx be^{\lambda_2 t}. \tag{11.61}$$

The particular integral A_* can be obtained by assuming the form $A = K$ where K is the undetermined coefficient, we have $dA/dt = 0$, $d^2 A/dt^2 = 0$, and

$$qK = rw, \tag{11.62}$$

or

$$K = \frac{rw}{q}. \tag{11.63}$$

Therefore, the general solution becomes

$$A = be^{\lambda_2 t} + \frac{rw}{q}. \tag{11.64}$$

The typical timescale τ for this system can be defined as $\tau = 1/|\lambda_2| = 1/0.0124 \approx 80$ years, so that the solution becomes

$$A = be^{-t/\tau} + \frac{rw}{q}. \tag{11.65}$$

Using the typical value of $r = 0.884/\text{year}$, and initial value $A(0) = 0$ ppm at $t = 0$, and $w \approx 0.014/\text{year} \times A_0 \approx 3.92$ ppm/year (the estimated level of emission in year 2000, equivalent to about 8.4 gigatons/year), we have

$$b \times e^0 + \frac{\alpha r}{q} = 0, \tag{11.66}$$

or

$$b = -\frac{rw}{q}.$$

The solution now becomes

$$A(t) = \frac{rw}{q}(1 - e^{-t/\tau}), \qquad (11.67)$$

or after plugging the numbers, we have

$$A(t) \approx 281.7(1 - e^{-t/80}). \qquad (11.68)$$

So the long-term excess of CO_2 will be 281.7 ppm, thus the actual CO_2 will be $281.7 + A_0 \approx 561.7$ ppm, which is about twice the pre-industrial concentration. Even with this constant rate, we can reach the maximum in 200 years. If the emission is increased to a higher level at 20 gigatons/year (in the year 2100), then the equilibrium CO_2 concentration will approach $670 + A_0 = 950$ ppm, and we can expect to reach to twice the pre-industrial level more quickly. This simple model shows that we have to reduce the CO_2 significantly over a long period.

11.5.2 Flexural Deflection of Lithosphere

The governing equation for the lithosphere deflection u under a load $P(x)$ can be written as

$$D\frac{d^4u}{dx^4} + g(\rho_m - \rho_w)u = P(x), \qquad (11.69)$$

where ρ_m, and ρ_w are the densities of the mantle and water, respectively. g is the acceleration due to gravity, and D is the flexural rigidity defined as

$$D = \frac{EH^3}{12(1 - \nu^2)}, \qquad (11.70)$$

where E and ν are Young's modulus and Poisson's ratio of the lithosphere, respectively. H is the thickness of the lithosphere (see Fig. 11.2). A special case of the load variation is periodic, that is

$$P(x) = \rho_s g h_0 \sin\frac{2\pi x}{\lambda}, \qquad (11.71)$$

where λ and h_0 are the typical wavelength and amplitude of the topological variations, respectively. ρ_s is the density of the load.

Now let us solve this equation and also provide the appropriate boundary conditions. The homogeneous equation becomes

$$Du'''' + g(\rho_m - \rho_w)u = 0, \qquad (11.72)$$

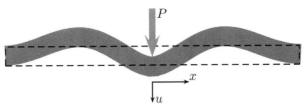

Figure 11.2: Deflection of the lithosphere as a thin shell.

whose characteristic equation (using $u = e^{\omega x}$) becomes

$$Dw^4 + g(\rho_m - \rho_w) = 0, \tag{11.73}$$

whose solution is

$$\omega = \left[\frac{-g(\rho_m - \rho_w)}{D}\right]^{\frac{1}{4}} = \sqrt[4]{-1}\sqrt[4]{\frac{\rho(\rho_m - \rho_w)}{D}}. \tag{11.74}$$

Since $\sqrt{-1} = i$ and

$$\sqrt[4]{-1} = \sqrt{i} = \frac{1}{\sqrt{2}}(\pm 1 \pm i), \tag{11.75}$$

and they correspond to four solutions

$$\omega_{1,2} = -(1 \pm i)\beta, \qquad \omega_{3,4} = (1 \pm i)\beta, \qquad \beta = \sqrt[4]{\frac{(\rho_m - \rho_w)g}{4D}}, \tag{11.76}$$

where we have absorbed the factor $1/\sqrt{2} = \sqrt[4]{1/4}$ into β. We have the complementary function

$$u = Ae^{(-\beta+i\beta)x} + Be^{(-\beta-i\beta)x} + Ce^{(\beta+i\beta)x} + De^{(\beta-i\beta)x}$$

$$= Ae^{-\beta x}[\cos\beta x + i\sin\beta x] + Be^{-\beta x}[\cos(-\beta x) + i\sin(-\beta x)]$$

$$+ Ce^{\beta x}[\cos\beta x + i\sin\beta x] + De^{\beta x}[\cos(-\beta x) + i\sin(-\beta x)]$$

$$= e^{-\beta x}[(A+B)\cos(\beta x) + i(A-B)\sin(\beta x)]$$

$$+ e^{\beta x}[(C+D)\cos(\beta x) + i(C-D)\sin(\beta x)] \tag{11.77}$$

where A and B are arbitrary constants. Here we have used $\sin(-\beta x) = -\sin(\beta x)$. This expression can be rewritten more compactly as

$$u = e^{-\beta x}[P\cos(\beta x) + Q\sin(\beta x)] + e^{\beta x}[U\cos(\beta x) + V\sin(\beta x)], \tag{11.78}$$

where $P = A + B$, $Q = i(A - B)$, $U = C + D$ and $V = i(C - D)$ are the new constants.

Since the amplitude of the deflection must be finite, the solutions $\omega_{3,4}$ are not possible because $e^{\beta x} \to \infty$ will grow exponential as $x \to \infty$. Therefore, $U = V = 0$. Only solutions $\omega_{1,2}$ are acceptable.

For the particular integral, let us try the form

$$u_* = u_0 \sin(\frac{2\pi x}{\lambda}). \tag{11.79}$$

After substituting into the original equation, we have

$$Du_0(\frac{2\pi}{\lambda})^4 \sin(\frac{2\pi x}{\lambda}) + g(\rho_m - \rho_w)u_0 \sin(\frac{2\pi x}{\lambda}) = \rho_s g h_0 \sin(\frac{2\pi x}{\lambda}).$$

After dividing both sides by $\sin(2\pi x/\lambda)$, we have

$$u_0[D(\frac{2\pi}{\lambda})^4 + g(\rho_m - \rho_w)] = \rho_s g h_0, \tag{11.80}$$

or

$$u_0 = \frac{\rho_s g h_0}{[D(\frac{2\pi}{\lambda})^4 + g(\rho_m - \rho_w)]} = \frac{h_0}{\frac{D}{\rho_s g}(\frac{2\pi}{\lambda})^4 + \frac{(\rho_m - \rho_w)}{\rho_s}}, \tag{11.81}$$

which is the well-known Turcotte-Schubert solution.

So the general solution now becomes

$$u = e^{-\beta x}[A\cos(\beta x) + B\sin(\beta x)]$$

$$+ \frac{h_0}{[\frac{D}{\rho_s g}(\frac{2\pi}{\lambda})^4 + \frac{(\rho_m - \rho_w)}{\rho_s}]} \sin\frac{2\pi x}{\lambda}. \tag{11.82}$$

The constants A and B should be determined by boundary conditions. If we are more interested in the deformation over a large scale, then x is large. So $\beta > 0$ implies that $e^{-\beta x} \to 0$ as $x \to \infty$. This means that the particular integral is more important in this case.

Here we can discuss two asymptotic cases even without determining A and B. From the above solution, we know that

$$\frac{D}{\rho_s g}(\frac{2\pi}{\lambda})^4 = 1, \tag{11.83}$$

defines a critical wave length λ_*, called the flexural wavelength

$$\lambda_* = 2\pi \sqrt[4]{\frac{D}{\rho_s g}}. \tag{11.84}$$

If $\lambda \gg 1$, we have

$$u_0 = \frac{h_0}{(\frac{\lambda_*}{\lambda})^4 + \frac{(\rho_m - \rho_w)}{\rho_s}} \approx \frac{h_0 \rho_s}{(\rho_m - \rho_w)}, \tag{11.85}$$

which is essentially the solution at the isostatic equilibrium.

On the other hand, if $\lambda \ll \lambda_*$, we have

$$u_0 = \frac{h_0}{(\frac{\lambda_*}{\lambda})^4 + \frac{(\rho_m - \rho_w)}{\rho_s}} \approx h_0 (\frac{\lambda}{\lambda_*})^4 \to 0. \qquad (11.86)$$

This means that the lithosphere is extremely rigid and the deflection on the short wavelength is negligible.

11.5.3   Glacial Isostatic Adjustment

The behaviour of solid and/or molten rocks will be very different if we look at them on different timescales. On a geological timescale, they are highly viscous. In fact, viscosity is important for many phenomena in earth sciences, including the mantle convection, post-glacier rebound and isostasy. The last glacial period ended about 10,000 to 15,000 years ago. During the ice age, thick glaciers caused the deformation of the surface of the crust, thus depleting the mantle materials or causing it to flow away. When the ice age ended, the retreat of the glacier would lead to the rebound or uplift of the depressed crust, accompanied by the flow back of the mantle materials. This process is similar to the process of dipping a finger briefly in soft or liquid chocolate. As the mantle is highly viscous with a viscosity of about 10^{21} Pa s, the rebound is a very slow process lasting several thousands to tens of thousands of years. This process is called post-glacial rebound, or glacial isostatic adjustment.

The uplift is typically a few millimetres a year with a relaxation time of a few thousand years for the post-glacial rebounds in north Europe and north America. Under the assumptions that the uplift process is similar to a viscoelastic relaxation process, the governing equation in the simplest case can be written as

$$\frac{\partial h}{\partial t} = -\frac{(h - h_0)}{\tau}, \qquad (11.87)$$

where h is the uplift, and h_0 is the total amount of uplift which will eventually be achieved as $t \to \infty$. τ is the characteristic timescale of relaxation or response timescale

$$\tau = \frac{4\pi\eta}{\rho_m g \lambda}, \qquad (11.88)$$

where η and ρ_m are the viscosity (or dynamic viscosity) and density of the mantle, respectively. g is the acceleration due to gravity, and λ is the typical wavelength of the uplift.

Equation (11.87) is a simple first-order differential equation, its solution can be obtained by direct integration if we use

$$z = h - h_0, \qquad (11.89)$$

which is the remaining uplift to be achieved. Since $\frac{dz}{dt} = \frac{dh}{dt}$, the governing equation (11.87) becomes

$$\frac{dz}{dt} = -\frac{z}{\tau}, \tag{11.90}$$

which leads to

$$\int \frac{1}{z} dz = -\int \frac{1}{\tau} dt, \quad \text{or} \quad \ln z = -\frac{t}{\tau} + A, \tag{11.91}$$

where A is the constant of integration. This gives

$$z = Be^{-t/\tau}, \tag{11.92}$$

where $B = e^A$. At $t = 0$, we have $h(0) = 0$ so that $z(0) = B = -h_0$. The solution now becomes

$$z = -h_0 e^{-t/\tau}, \tag{11.93}$$

and the uplift is

$$h(t) = z + h_0 = h_0(1 - e^{-t/\tau}). \tag{11.94}$$

Therefore, we have $h \to h_0$ as $t \to \infty$ as expected.

From field observations, we now know that the relaxation time for the Fennoscandian uplift in north Europe is about 4900 to 6750 years. Let us see how equation (11.88) fits. From the typical values of $\eta = 10^{21}$ Pa s, $\rho_m = 3300$ kg/m^3, $\lambda = 2000$ km$= 2 \times 10^6$ meters, and $g = 9.8$ m/s^2, we have

$$\tau = \frac{4\pi\eta}{\rho g \lambda}$$

$$= \frac{4\pi \times 10^{21}}{3300 \times 9.8 \times 2 \times 10^6} \approx 1.94 \times 10^{11} \text{ seconds} \approx 6160 \text{ years}, \tag{11.95}$$

where we have used 1 year$= 3.15 \times 10^7$ seconds. This is about right.

Conversely, if we know the relaxation time τ, we can estimate the viscosity of the mantle. If we use the average $\bar{\tau} \approx 5500$ years$= 1.73 \times 10^{11}$ seconds, we have

$$\eta = \frac{\bar{\tau}\rho g \lambda}{4\pi}$$

$$\approx \frac{1.73 \times 10^{11} \times 3300 \times 9.8 \times 2 \times 10^6}{4\pi} \approx 0.9 \times 10^{21} \text{ Pa s}, \tag{11.96}$$

which is a better estimate than 10^{21} Pa s.

In addition, we could measure the uplifts h_1 and h_2 at two different times t_1 and t_2, respectively, we can calculate the ultimate uplift h_0. From any two observations

$$h_1 = h_0(1 - e^{-t_1/\tau}), \tag{11.97}$$

and

$$h_2 = h_0(1 - e^{-t_2/\tau}), \tag{11.98}$$

we have

$$h_2 - h_1 = h_0(e^{-t_1/\tau} - e^{-t_2/\tau}), \tag{11.99}$$

or

$$h_0 = \frac{(h_2 - h_1)}{(e^{-t_1/\tau} - e^{-t_2\tau})}. \tag{11.100}$$

If we take the ratio of (11.97) to (11.98), we have

$$\frac{h_1}{h_2} = \frac{1 - e^{-t_1/\tau}}{1 - e^{-t_2/\tau}}, \tag{11.101}$$

which can be solved to get τ by numerical methods such as the Newton-Raphson method, to be introduced in the final chapter. Mathematically, this is achievable, but in practice, it is very difficult to determine t_1 and t_2 accurately. However, we can indeed measure the difference $\Delta t = t_2 - t_1$ accurately. In this case, the better alternative is to use the rate $r = dh/dt$ of the uplift rather than the uplift itself.

By differentiating the solution (11.94), we have

$$\frac{dh}{dt} = \frac{h_0}{\tau}e^{-t/\tau}. \tag{11.102}$$

Again two observations of rates r_1 and r_2 at two different times t_1 and t_2, respectively, we have

$$r_1 = \frac{h_0}{\tau}e^{-t_1/\tau}, \qquad r_2 = \frac{h_0}{\tau}e^{-t_2/\tau}, \tag{11.103}$$

whose ratio becomes

$$\frac{r_1}{r_2} = \frac{e^{-t_1/\tau}}{e^{-t_2/\tau}} = e^{(t_2-t_1)/\tau} = e^{\Delta t/\tau}. \tag{11.104}$$

Taking logarithms and rearranging the equation, we have

$$\tau = \frac{\Delta t}{\ln(r_1/r_2)} = \frac{(t_2 - t_1)}{\ln(r_1/r_2)}. \tag{11.105}$$

This is a more useful formula to determine the response timescale.

Alternatively, if we know τ and the current uplift rate, we can predict the uplift in the future. We know the current uplift rate in the year 2000 in north Europe is about $r_{2000} = 0.9$ cm/year. If we use $\tau = 5500$ years, we can estimate the uplift rate in the year 3000. Therefore, $\Delta t = 1000$ years. From (11.104), we have

$$\frac{r_{2000}}{r_{3000}} = e^{\Delta t/\tau} = e^{1000/5500} \approx 1.199, \tag{11.106}$$

which means

$$r_{3000} = \frac{r_{2000}}{1.199} \approx \frac{0.9}{1.199} \approx 0.75 \text{ cm/year}. \tag{11.107}$$

Chapter 12

Partial Differential Equations

Ordinary differential equations are sometimes complicated enough, but their applications are often limited as they only involve a single independent variable. In earth sciences, quantities usually depend on many other variables, and the governing equations for many geophysical and geological processes are often expressed in terms of partial differential equations (PDEs). PDEs are much more complicated than ordinary differential equations. Here we will briefly introduce the fundamentals of PDEs and focus on the linear heat conduction and diffusion equation which are of importance in earth sciences.

While introducing the basics of PDEs, we will focus on the application of PDEs to the heat transfer process modelling, outlining Kelvin's estimate of the age of the Earth, and the general solution of the cooling of the lithosphere.

12.1 Introduction

A partial differential equation is a relationship or equation containing one or more partial derivatives. Similar to the ordinary differential equation, the highest nth partial derivative is referred to as the order n of the partial differential equation. In general, we can write it as a generic function ϕ

$$\phi(u, x, y, \frac{\partial u}{\partial x}, \frac{\partial u}{\partial y}, \frac{\partial^2 u}{\partial x^2}, \frac{\partial^2 u}{\partial y^2}, \frac{\partial^2 u}{\partial x \partial y}, ...) = 0, \qquad (12.1)$$

where u is the dependent variable, and $x, y, ...$ are the independent variables.

For example, the one-dimensional wave equation

$$\frac{\partial^2 u}{\partial t^2} = v^2 \frac{\partial^2 u}{\partial x^2}, \tag{12.2}$$

is a second-order partial differential equation. Here time t and space x are two independent variables, and u is the displacement. The constant v is the wave speed. In many mathematical books, compact subscript forms are often used to denote partial derivatives for simplicity and clarity. That is $u_x \equiv \frac{\partial u}{\partial x}$, $u_{xx} \equiv \frac{\partial^2 u}{\partial x^2}$, $u_y \equiv \frac{\partial u}{\partial y}$, $u_{xy} \equiv \frac{\partial^2 u}{\partial x \partial y}$, and so on and so forth. Using such compact notations, the wave equation can be written as

$$u_{tt} = v^2 u_{xx}. \tag{12.3}$$

12.2 Classic Equations

There are different types of PDEs, and the three basic types for two independent variables are based on the following general linear equation

$$a\frac{\partial^2 u}{\partial x^2} + b\frac{\partial^2 u}{\partial x \partial y} + c\frac{\partial^2 u}{\partial y^2} = f(x, y, u, u_x, u_y), \tag{12.4}$$

where a, b and c are known functions of x and y. In the simplest cases, they can all be constants. Similar to the case of quadratic equations, we define $\Delta = b^2 - 4ac$.

If $\Delta > 0$, the partial differential equation is said to be hyperbolic. The wave equation

$$\frac{\partial^2 u}{\partial t^2} = v^2 \frac{\partial^2 u}{\partial x^2}, \tag{12.5}$$

is a good example if we let $t = y$, $a = v^2$, $b = 0$ and $c = -1$. This means $\Delta = 4v^2 > 0$ and the equation is hyperbolic.

If $\Delta < 0$, the PDE is said to be elliptic. The Laplace equation

$$\frac{\partial^2 u}{\partial x^2} + \frac{\partial^2 u}{\partial y^2} = 0, \tag{12.6}$$

is elliptic because $\Delta = -4 < 0$ due to $a = c = 1$ and $b = 0$.

If $\Delta = 0$, the PDE is said to be parabolic. The heat conduction equation

$$\frac{\partial u}{\partial t} = \kappa \frac{\partial^2 u}{\partial x^2}, \tag{12.7}$$

is parabolic as $a = \kappa$, $c = 0$ for $t = y$ and $b = 0$ so that $\Delta = 0$.

These three types are the most commonly used in mathematical modelling. There are other types such as the mixed type, as the coefficient functions depend on x and y, and in different regions, they may take different values and the sign of Δ will change, and subsequently the type of the equation will also change.

The classification of PDEs will usually provide some information about how the partial differential equation behaves, and we can expect certain behaviour and characteristics of the solution of our interest. For example, the wave equation, as the name suggests, will support certain travelling waves such as the vibrations of an elastic string, and the acoustic vibrations of the air. In fact, almost all hyperbolic equations will support some forms of waves under appropriate conditions. On the other hand, for the parabolic equations when $\Delta = 0$, the solution will be smooth and any initial irregularity in the profile will be smoothed out. This is typically the behaviour of diffusion and heat conduction. For example, the sugar concentration in a cup of coffee will tend to become uniform after some time, and the temperature of a swimming pool will tend to be uniform to some degree. In the rest of this chapter, we will discuss the heat conduction equation in more detail as it is very important in earth sciences.

There are many methods for solving partial differential equations, including separation of variables, Fourier transforms, Laplace transform, Green's functions, similarity variables, spectral methods, and others. In the rest of this chapter, we will only introduce the solution of heat conduction equations and its application in the modelling of the cooling process of lithosphere.

12.3   Heat Conduction

Suppose we are interested in the cooling of the lithosphere. In this case, we are only interested in the one-dimensional case of how the temperature T varies with time t and the depth z (see Fig. 12.1). We know that the governing equation is the classic heat conduction

$$\frac{\partial T}{\partial t} = \kappa \frac{\partial^2 T}{\partial z^2}, \qquad (12.8)$$

where $\kappa = K/\rho c_p$ is the thermal diffusivity, and K is the heat conductivity. ρ is the density of the lithosphere and c_p is the specific heat capacity. The domain of our interest consists of $t \geq 0$ and $0 \leq z < \infty$.

The boundary condition at $z = 0$ is that the temperature is constant. That is, $T = T_s$ at $z = 0$. The temperature at the other boundary where $z \to \infty$ should approach the temperature T_m of the underlying mantle. In this problem, we can assume that the initial temperature is $T(z, t = 0) = T_s$. We know that κ has a unit of m^2/s, so κt has a unit of m^2. This means that $\sqrt{\kappa t}$ defines a length as its unit is m. Therefore, the ratio $z/\sqrt{\kappa t}$ is dimensionless, which is convenient

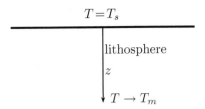

Figure 12.1: Heat conduction in 1D.

for mathematical analysis. In fact, we can define a similarity variable

$$\zeta = \frac{z}{\sqrt{4\kappa t}}, \tag{12.9}$$

where the factor 4 is purely for convenience to avoid a factor $1/2$ appearing everywhere in the end equation. Now we have $T(z,t) = f(\zeta)$. Using the chain rules of differentiations

$$\frac{\partial f}{\partial z} = \frac{\partial f}{\partial \zeta}\frac{\partial \zeta}{\partial z} = \frac{1}{\sqrt{4\kappa t}}\frac{\partial f}{\partial \zeta}, \quad \frac{\partial^2 f}{\partial z^2} = \frac{\partial \zeta}{\partial z}\frac{\partial \zeta}{\partial z}\frac{\partial^2 f}{\partial \zeta^2} = (\frac{1}{\sqrt{4\kappa t}})^2\frac{\partial^2 f}{\partial \zeta^2} = \frac{1}{4\kappa t}\frac{\partial^2 f}{\partial \zeta^2},$$

and

$$\frac{\partial f}{\partial t} = \frac{\partial f}{\partial \zeta}\frac{\partial \zeta}{\partial t} = \frac{z}{\sqrt{4\kappa}}(-\frac{1}{2}t^{-3/2})\frac{\partial f}{\partial \zeta} = \frac{-z}{2t\sqrt{4\kappa t}}\frac{\partial f}{\partial \zeta} = -\frac{\zeta}{2t}\frac{\partial f}{\partial \zeta}, \tag{12.10}$$

we can write the original PDE (12.8) as

$$-\frac{\zeta}{2t}f' = \frac{\kappa}{4\kappa t}f'', \tag{12.11}$$

where we have used $f' = \frac{df}{d\zeta} = \frac{\partial f}{\partial \zeta}$ and $f'' = \frac{d^2 f}{d\zeta^2} = \frac{\partial^2 f}{\partial \zeta^2}$ because $f(\zeta)$ depends on ζ only, and the partial derivatives and the standard derivatives are the same.

The above equation can be rearranged as

$$\frac{f''}{f'} = -2\zeta. \tag{12.12}$$

The chain rule of differentiation implies that $(\ln y(x))' = (1/y(x))y'(x)$. Replacing $y(x)$ by f' or letting $y(x) = f'$, we have $(\ln f')' = f''/f'$. Integrating the above equation once, we get

$$\ln f' = -\zeta^2 + C, \tag{12.13}$$

where C is the constant of integration. This again can be written as

$$f' = Ke^{-\zeta^2}, \tag{12.14}$$

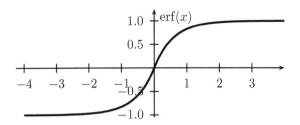

Figure 12.2: Plot of the error function $\text{erf}(x)$.

where $K = e^C$ is a constant. Integrating this with respect to ζ, we have

$$T = f(\zeta) = A\text{erf}(\frac{z}{\sqrt{4\kappa t}}) + B, \qquad (12.15)$$

where $A = K\sqrt{\pi}/2$. Here A and B are two undetermined constants. The error function is defined by

$$\text{erf}(x) = \frac{2}{\sqrt{\pi}} \int_0^x e^{-\zeta^2} d\zeta. \qquad (12.16)$$

From its definition, we know that $\text{erf}(0) = 0$. At infinity, we have

$$\text{erf}(\infty) \to +1, \quad \text{as } x \to +\infty, \quad \text{erf}(\infty) \to -1, \quad \text{as } x \to -\infty. \quad (12.17)$$

We know these are true because $\int_0^\infty e^{-x^2} dx = \frac{\sqrt{\pi}}{2}$.
Using the boundary condition $T = T_s$ at $z = 0$, we have

$$A \, \text{erf}(0) + B = T_s, \qquad (12.18)$$

which gives $B = T_s$. From the far field boundary $T = T_m$ as $z \to \infty$, we have

$$A \, \text{erf}(\infty) + B = A + B = T_m. \qquad (12.19)$$

In combination with $B = T_s$, we have $A = T_m - T_s$. Therefore, the final solution for the temperature profile becomes

$$T(z, t) = T_s + (T_m - T_s)\text{erf}(\frac{z}{\sqrt{4\kappa t}}). \qquad (12.20)$$

Example 12.1: William Thomson (later Lord Kelvin) was the first to estimate the age of the Earth from physical principles. By assuming that the Earth is a half-space with the condition that the temperature tends to T_0 at infinite depth $z \to \infty$, the solution of the temperature can be written

$$T = T_0\text{erf}\left(\frac{z}{2\sqrt{\kappa t}}\right),$$

where z is the depth, and erf(ζ) is the error function. Using $\zeta = \frac{z}{2\sqrt{\kappa t}}$ and $dT/d\zeta = d[\frac{2T_0}{\sqrt{\pi}} \int_0^\zeta e^{-\zeta^2} d\zeta]/d\zeta = \frac{2T_0}{\sqrt{\pi}} e^{-\zeta^2}$ [see (12.23)], the thermal gradient can be calculated by

$$G = \frac{\partial T}{\partial z} = \frac{\partial T}{\partial \zeta}\frac{\partial \zeta}{\partial z} = \frac{2T_0}{\sqrt{\pi}} e^{-\zeta^2} \frac{1}{2\sqrt{\kappa t}} = \frac{T_0}{\sqrt{\pi \kappa t}} e^{-\zeta^2}.$$

At the Earth's surface $z = 0$ and $\zeta = 0$, we have $G = \frac{T_0}{\sqrt{\pi \kappa t}}$, and the heat flux $Q = -K\frac{\partial T}{\partial z}\big|_{z=0} = -\frac{KT_0}{\sqrt{\pi \kappa t}}$. We know the observed heat flux Q or temperature gradient today; the age of the Earth is

$$t \approx \frac{(T_0/G)^2}{\pi \kappa}.$$

Kelvin in 1863 used the measured values at that time $\kappa \approx 1.2 \times 10^{-6}$ m^2/s, $T_0 = 3900°$C, and $G = 36°$C/km$=0.036°$C/m, he estimated that

$$t = \frac{(3900/0.036)^2}{\pi \times 1.2 \times 10^{-6}} \approx 3.1 \times 10^{15} \text{ seconds} \approx 98.7 \text{ Ma},$$

where we have used 1 Ma $= 1,000,000 \times 365 \times 24 \times 3600 = 3.15 \times 10^{13}$ seconds. Taking account of the uncertainties of these measurements, Kelvin set a limit of 24 Ma to 400 Ma on the age of the Earth. Obviously, this greatly underestimated the age of the Earth, but such calculations were a significant progress at that time.

- -

12.4　Cooling of the Lithosphere

Now we investigate further in modelling the cooling process of the lithosphere near the mid-ocean ridge (see Fig. 12.3).

The solution to a heat conduction problem in a semi-infinite domain usually takes the form

$$T = T_s + (T_m - T_s)\text{erf}\left(\frac{z}{2\sqrt{\kappa t}}\right), \tag{12.21}$$

where T_s is the temperature at $z = 0$, and T_m is the temperature of the mantle.

The thermal gradient at $z = 0$ can be calculated by

$$q = -K\frac{\partial T}{\partial z} = -K\frac{\partial T}{\partial \zeta}\frac{\partial \zeta}{\partial z}. \tag{12.22}$$

Since $\frac{\partial \zeta}{\partial z} = \frac{1}{2\sqrt{\kappa t}}$ and

$$\frac{\partial \text{erf}(\zeta)}{\partial \zeta} = \frac{2}{\sqrt{\pi}}\frac{\partial}{\partial \zeta}\left[\int_0^\zeta e^{-u^2} du\right] = \frac{2}{\sqrt{\pi}} e^{-\zeta^2}, \tag{12.23}$$

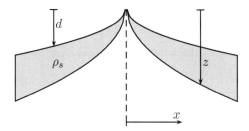

Figure 12.3: Cooling and depth variations of the lithosphere near the mid-ocean ridge.

we have

$$q = -\frac{K(T_m - T_s)}{\sqrt{\pi\kappa t}}e^{-\zeta^2} = -\frac{K(T_m - T_s)}{\sqrt{\pi\kappa t}}e^{-z^2/4\kappa t}. \tag{12.24}$$

The negative sign highlights the fact that the heat flow is upwards in the opposite direction of z. We are concerned with the value at $z = 0$, therefore we have $q_0 = q\big|_{z=0} = -\frac{K(T_m - T_s)}{\sqrt{\pi\kappa t}}$.

For practical application, we can set $T_s = 0$ for the Earth if we use T in degrees, not in terms of the absolute temperature. Then, we have

$$T = T_m\mathrm{erf}\left(\frac{z}{2\sqrt{\kappa t}}\right), \qquad q_0 = -\frac{KT_m}{\sqrt{\pi\kappa t}}. \tag{12.25}$$

For a given time t, we can estimate the heat flux. Conversely, if we know the value of the heat flux, we can estimate the age of the lithosphere. This equation is also the basis for the first estimation of the age of the Earth by Lord Kelvin. In addition, we can also estimate the temperature T_m once we know the age and the heat flux.

It is observed that the depths of the ocean near the mid-ocean ridge vary with time and also the distance from the ridge. The depth variations $d(t)$ relative to the depth of the mid-ocean ridge d_0 are adjusted according to the isostatic balance between the mass deficiency caused by $(\rho_m - \rho_w)gd$ and the density increase by the cooling of the lithosphere at any $z = L$ and $L \to \infty$. Here ρ_m and ρ_w are the densities of the mantle and water, respectively. That is

$$(\rho_m - \rho_w)g(d - d_0) + \int_0^L \alpha\rho_m(T - T_m)gdz = 0, \tag{12.26}$$

where α is the coefficient of thermal expansion. Using (12.21), we have

$$d = d_0 + \frac{\rho_m\alpha}{(\rho_m - \rho_w)}\int_0^L (T_m - T_s)\mathrm{erfc}\left(\frac{z}{2\sqrt{\kappa t}}\right)dz = d_0 + \Lambda\int_0^L \mathrm{erfc}\left(\frac{z}{2\sqrt{\kappa t}}\right)dz,$$

where $\Lambda = \frac{\rho_m \alpha (T_m - T_s)}{(\rho_m - \rho_w)}$, and erfc is the complementary error function, defined as $\text{erfc}(\zeta) = 1 - \text{erf}(\zeta)$. In order to integrate $\int \text{erfc}\left(\frac{z}{2\sqrt{\kappa t}}\right) dz$, we now use the integration by parts for the case of $L \to \infty$

$$I = \int_0^\infty \text{erfc}\left(\frac{z}{2\sqrt{\kappa t}}\right) dz.$$

Let $v' = 1$ and $u = \text{erfc}\left(\frac{z}{2\sqrt{\kappa t}}\right)$, we have $v = z$ and

$$\frac{du}{dz} = \frac{\partial \text{erfc}\left(\frac{z}{2\sqrt{\kappa t}}\right)}{\partial z} = \frac{\partial [1 - \text{erf}\left(\frac{z}{2\sqrt{\kappa t}}\right)]}{\partial z} = -\frac{1}{\sqrt{\pi \kappa t}} e^{-\zeta^2}. \qquad (12.27)$$

We then have

$$I = \int_0^\infty \text{erfc}\left(\frac{z}{2\sqrt{\kappa t}}\right) dz = \left[z \text{erfc}\left(\frac{z}{2\sqrt{\kappa t}}\right) \right]_0^\infty - \left(-\frac{1}{\sqrt{\pi \kappa t}}\right) \int_0^\infty z e^{-\zeta^2} dz.$$

Since $\text{erfc}(\infty) = 1 - \text{erf}(\infty) = 0$, $z/2\sqrt{\kappa t} = \zeta$ and $d\zeta = dz/2\sqrt{\kappa t}$, we have

$$I = 0 + \frac{4\sqrt{\kappa t}}{\sqrt{\pi}} \int_0^\infty \zeta e^{-\zeta^2} d\zeta. \qquad (12.28)$$

Using the integral

$$\int_0^\infty y e^{-y^2} dy = \frac{1}{2} \int_0^\infty e^{-y^2} d(y^2) = \frac{1}{2}\left[-e^{-y^2} \right]_0^\infty = \frac{1}{2}, \qquad (12.29)$$

we have

$$I = \frac{4\sqrt{\kappa t}}{\sqrt{\pi}} \times \frac{1}{2} = \frac{2\sqrt{\kappa t}}{\sqrt{\pi}}. \qquad (12.30)$$

Therefore, we finally have the solution

$$d = d_0 + \left(2\Lambda\sqrt{\frac{\kappa t}{\pi}}\right) = d_0 + \frac{2\alpha\rho_m(T_m - T_s)}{(\rho_m - \rho_w)}\sqrt{\frac{\kappa t}{\pi}}. \qquad (12.31)$$

This is the Parsons-Sclater formula for bathymetry. Using the typical values of $T_s = 0°C$, $\kappa = 8 \times 10^{-7}$ m^2/s, $T_m = 1350°C$, $\alpha = 3 \times 10^{-5}$/C, $\rho_m = 3000$ kg/m^3, $\rho_w = 1025$ kg/m^3, and $d_0 = 2500$ m, we have

$$d = 2500 + \frac{2 \times 3 \times 10^{-5} \times 3000 \times (1350 - 0)}{(3000 - 1025)}\sqrt{\frac{8 \times 10^{-7}}{\pi}}\sqrt{t}$$

$$= 2500 + 6.209 \times 10^{-5}\sqrt{t}, \qquad (12.32)$$

where t is in seconds. If we convert it to the unit of million years, we have to multiply a factor of $365 \times 24 \times 3600 \times 10^6 = 3.15 \times 10^{13}$. Therefore, the final formula becomes

$$d = 2500 + 6.209 \times 10^{-5}\sqrt{3.15 \times 10^{13}t} \approx 2500 + 348.5\sqrt{t}, \qquad (12.33)$$

where t is in Ma. This formula is almost the same as the best fit to bathymetry data $d = 2500 + 350\sqrt{t}$ for $t < 70$ Ma.

Chapter 13

Geostatistics

Statistics, especially geostatistics is an important toolset for earth sciences. It concerns the data collection and interpretation, analysis and characterisation of numerical data and design and analysis of sampling.

In this chapter, we will briefly introduce the fundamentals of statistics. We will then apply the introduced techniques to study the propagation of errors, linear regression of earthquake data, and Brownian motion including diffusion process.

13.1 Random Variables

13.1.1 Probability

When we conduct an experiment such as tossing a coin or recording the outside noise level, we are dealing with the outcomes of the experiment. For the tossing of a coin, there are only two possible outcomes: heads (1) or tails (0). If we toss 2 coins at the same time, we want to see how many heads may appear. If we use $X_i (i = 1, 2)$ to represent the values of the outcomes of the two coins, we have

$$X_1 + X_2, \tag{13.1}$$

to represent the number of heads. Here X_i are the random variables. Since they only take discrete values 0 and 1, they are also called discrete random variables. Other examples of discrete random variables include the number of cars passing through a junction in a day, the number of phone calls you received in a week, the number of major earthquakes in the world in a year, and the number of rock samples you collected during a field trip.

On the other hand, the noise level Y can take any values within a certain range. This random variable is called a continuous random variable. Other continuous random variables are the temperature variations in a year, the stress variations in an earthquake zone, and the velocity of a drifting plate.

The set of all the possible outcomes forms the sample space, and the elements of a sample space are called outcomes. Each subset of a sample space is called an event. For example, rolling a die has 6 possible outcomes, so the sample space is simply the set of all the possible outcomes: $\Omega = \{1, 2, 3, 4, 5, 6\}$. An event such as $A = \{2, 3, 5\}$ is the event that corresponds to the case when the face values of the die are prime numbers.

Probability P is a number or an expected frequency assigned to an event A that indicates how likely it is that the event will occur when a random experiment is performed. In order to show the probability P is associated with event A, we usually write this probability as $P(A)$. If we conduct a large number of fair trials such as tossing a coin 5,000 times, the probability of an event such as heads can be calculated by

$$P(A) = \frac{N_A \text{ (number of outcomes in the event A)}}{N_\Omega \text{ (total number of outcomes)}}. \tag{13.2}$$

The total sum of all the probabilities of all the possible outcomes must be 1. For example, if we toss a coin 5000 times, the heads appear 2507 times. The probability of heads is $P(H) = 2507/5000 = 0.5014$. As there must be $(5000 - 2507) = 2493$ tails appear, the probability of tails is $P(T) = 2493/5000 = 0.4986$. The total probabilities are $P(H) + P(T) = 0.5014 + 0.4986 = 1.0$.

Two events are said to be independent if one event is not affected by whether or not the other event occurs. In this case, the probability P of both events occurring is the product of individual probabilities

$$P = P(A)P(B), \tag{13.3}$$

where $P(A)$ and $P(B)$ are the probabilities for events A and B, respectively. For example, if we toss a coin and roll a die at the same time, what is the probability that both a head and the number 5 occur? Let A be the event of a head, and B be the event of number 5 as the outcome of rolling a die. We have $P(A) = 1/2$ and $P(B) = 1/6$, therefore we have the probability of both events occurring at the same time $P = P(A)P(B) = (1/2) \times (1/6) = 1/12$.

Each value or outcome of a random variable may occur with certain probability, and the probability may vary. In this case, we use a probability density function $p(x)$ to represent how the probability varies with x. If there are n possible outcomes x_i for a discrete random variable, we have

$$\sum_{i=1}^{n} p(x_i) = 1. \tag{13.4}$$

The function $p(x)$ is also known as the distribution function. When the variable is discrete, the distribution is called discrete distribution.

If a random variable is continuous, the probability function $f(x)$ is usually defined for a range of values $x \in [a, b]$. The probability is often associated with an interval (rather than a distinct value) $(x, x + dx]$ or the random variable takes the value $x < X \le x + dx$ where dx is a small increment. The convention we used here is that we use the Roman capital letter to denote the random variable and its corresponding small lower case to denote its values.

The probability function for a continuous random variable is also called the probability density function (pdf). In this case, all the probabilities should be added to unity. That is

$$\int_a^b p(x)dx = 1. \tag{13.5}$$

A good example of discrete distribution functions is the Poisson distribution

$$p(k; \lambda) = \frac{\lambda^k e^{-\lambda}}{k!}, \qquad (\lambda > 0), \tag{13.6}$$

where $k = 0, 1, 2, 3, ...$ and λ is the parameter of the distribution. An excellent example of continuous distribution functions is the Gaussian or normal distribution

$$p(x) = \frac{1}{\sigma\sqrt{2\pi}} e^{-\frac{(x-\mu)^2}{2\sigma^2}}, \qquad -\infty < x < \infty \tag{13.7}$$

where x is continuous. μ and σ are two parameters. The normal distribution with $\mu = 0$ and $\sigma = 1$ is called the standard normal distribution.

13.1.2   Mean and Variance

Two main measures for a random variable X with a given probability distribution function $p(x)$ are its mean and variance.

The mean μ is also called the expected value of the distribution, denoted by $E[X]$ and is defined by

$$\mu \equiv E(X) = \int xp(x)dx, \tag{13.8}$$

for a continuous distribution and the integration is within the integration limits. Other notations such as $<X> \equiv \overline{X}$ are also interchangeable with μ. If the random variable is discrete, then the integration becomes the summation

$$\mu = \sum_i x_i p(x_i). \tag{13.9}$$

The variance, denoted by $\sigma^2 \equiv \mathrm{var}[X]$ is the mean of the deviation squared $(X - \mu)^2$. That is

$$\sigma^2 \equiv \mathrm{var}[X] = E[(X - \mu)^2] = \int_\Omega (x - \mu)^2 p(x)dx, \qquad (13.10)$$

where Ω is the sample space. The square root of the variance $\sigma = \sqrt{\mathrm{var}[X]}$ is called the standard deviation, which is denoted by the symbol σ. Obviously, for a discrete distribution, the variance simply becomes a sum

$$\sigma^2 = \sum_i (x - \mu)^2 p(x_i). \qquad (13.11)$$

The mean and variance of the Poisson distribution are both λ.

Example 13.1: Let us now try to calculate the mean and variance of the normal distribution

$$p(x) = \frac{1}{\sigma\sqrt{2\pi}} e^{-\frac{(x-\mu)^2}{2\sigma^2}}, \qquad -\infty < x < +\infty.$$

The mean is given by

$$E[X] = \int_{-\infty}^{\infty} xp(x)dx.$$

By setting $u = x - \mu$, we have $x = u + \mu$ and $du = dx$. In addition, the infinite limits of the integration remain the same by shifting any finite amount μ. The above equation becomes

$$E[X] = \frac{1}{\sigma\sqrt{2\pi}} \int_{-\infty}^{\infty} (u + \mu)e^{-u^2/2\sigma^2} du = \frac{1}{\sigma\sqrt{2\pi}}(I_1 + \mu I_2),$$

where

$$I_1 = \int_{-\infty}^{\infty} ue^{-u^2/2\sigma^2} du, \qquad I_2 = \int_{-\infty}^{\infty} e^{-u^2/2\sigma^2} du.$$

Using

$$\mathrm{erf}(x) = \frac{2}{\sqrt{\pi}} \int_0^x e^{-u^2} du,$$

and letting $v = u/\sqrt{2}\sigma$, we have

$$I_2 = \sigma\sqrt{2} \int_{-\infty}^{\infty} e^{-v^2} dv = \sigma\sqrt{\pi/2}\,\mathrm{erf}(v)\Big|_{-\infty}^{\infty} = \sigma\sqrt{2\pi},$$

where we have used $\mathrm{erf}(\infty) = 1$ and $\mathrm{erf}(-\infty) = -1$.
    Similarly, from $udu = \frac{1}{2}d(u^2)$, we have

$$I_1 = \int_{-\infty}^{\infty} ue^{-u^2/2\sigma^2} du = \frac{1}{2} \int_{-\infty}^{\infty} e^{-u^2/2\sigma^2} d(u^2)$$

$$= \sigma \int_{-\infty}^{\infty} e^{-v^2} dv^2 = \sigma e^{-v^2} \Big|_{-\infty}^{\infty} = 0.$$

Therefore, the mean becomes

$$E[X] = \frac{\mu I_2}{\sigma \sqrt{2\pi}} = \mu.$$

The variance is given by $\mathrm{var}[X]$ or

$$E[(x - \mu)^2] = \int_{-\infty}^{\infty} (x - \mu)^2 p(x) dx = \frac{1}{\sigma \sqrt{2\pi}} \int_{-\infty}^{\infty} (x - \mu)^2 e^{-(x-\mu)^2/2\sigma^2} dx.$$

Setting $w = (x - \mu)/\sqrt{2}\sigma$, we have $dx = \sigma\sqrt{2}dw$, and we have

$$\mathrm{var}[X] = \frac{\sqrt{2}\sigma 2\sigma^2}{\sigma \sqrt{2\pi}} \int_{-\infty}^{\infty} w^2 e^{-w^2} dw = \frac{2\sigma^2}{\sqrt{\pi}} \int_{-\infty}^{\infty} w^2 e^{-w^2} dw.$$

Using the result given in an earlier example in the chapter on integration

$$\int_{-\infty}^{\infty} w^2 e^{-w^2} dw = \frac{\sqrt{\pi}}{2},$$

we have

$$\mathrm{var}[X] = \frac{2\sigma^2}{\sqrt{\pi}} \frac{\sqrt{\pi}}{2} = \sigma^2.$$

This proves that the mean of the normal distribution is μ and the variance is σ^2 so that the standard deviation is σ.

- -

If we know the distribution function, it is always possible to calculate its mean and variance, though the integration is not an easy task. In earth sciences, we have often collected a lot of data, but we do not know exactly what the distribution is. In this case, we have to deal with sample mean and sample variance.

13.2 Sample Mean and Variance

If a sample consists of n independent observations x_1, x_2, ..., x_n on a random variable X such as the noise level, the sample mean can be calculated by

$$\bar{x} = \frac{1}{n} \sum_{i=1}^{n} x_i = \frac{1}{n}(x_1 + x_2 + ... + x_n), \tag{13.12}$$

which is the arithmetic average of the values of the data x_i.

The sample variance is defined by

$$S^2 = \frac{1}{n-1} \sum_{i=1}^{n} (x_i - \bar{x})^2. \tag{13.13}$$

It is worth pointing out that the factor is $1/(n-1)$ not $1/n$ because only $1/(n-1)$ will give the correct and unbiased estimate of the variance. The other way to think about this factor is that we need at least one value to estimate the sample mean, and we need at least two values to estimate the variance. Therefore, for n observations, only $(n-1)$ different values of variance can be obtained to estimate the total sample variance.

Example 13.2:   The average drifting velocity of the tectonic plate can be estimated using the data from Hawaiian Islands, formed as a result of a series of volcanic eruptions by a giant hot spot. The distance data are as follows: Kauai to Oahu is about 180 km with an age difference of 1.9 Ma (million years); Oahu to Molakai is 120 km with an age difference of 1.27 Ma; Molakai to Maui is 80 km with an age difference of 0.7 Ma; and Maui to Hawaii is 125 km with an age difference of 1 Ma. So the velocity of the plate movement from Kauai to Oahu is

$$v = \frac{d}{t} = \frac{180 \times 1000}{1.9 \times 10^6} \text{ m/year} \approx 0.095 \text{ m/year} = 9.5 \text{ cm/year}.$$

Similarly, other velocities are 9.4, 11.4, 12.5 cm/year.
      The mean velocity of this plate movement is

$$\bar{v} = \frac{1}{4}(9.5 + 9.4 + 11.4 + 12.5) = 10.7 \text{cm/year}.$$

The variance of the velocity (for $n = 4$) is

$$\sigma^2 = \text{var}[v] = \frac{1}{4-1}[(9.5 - 10.7)^2 + (9.4 - 10.7)^2$$

$$+(11.4 - 10.7)^2 + (12.5 - 10.7)^2] \approx \frac{1}{3} \times 6.86 \approx 2.3.$$

Therefore, the standard deviation is $\sigma = \sqrt{2.3} \approx 1.5$ cm/year.

13.3   Propagation of Errors

Any experiment will have measurement errors or uncertainty. There are two main type of errors: system errors, and random errors. System errors are errors that are systematic, either known or unknown, which will affect the measurement or observations in a systematic way. System

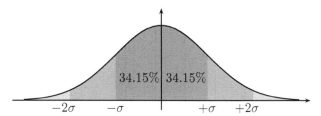

Figure 13.1: Standard deviation and confidence interval.

errors can be reduced by good design of experiments and analysis of contributing factors. Random errors are undeterministic, which are unavoidable but can be minimised by using higher-precision equipment.

The words 'errors' and 'uncertainty' are almost synonymic and thus interchangeable. However, we often use uncertainty to chacterise the variations of errors, while we use errors to measure the difference between the measured and the true value. Since the true value is rarely known, the errors are not easy to estimate. There are two ways to define errors: absolute error and relative error.

The absolute error is simply the difference between the observed and the true value or expected value. For example, we know that density of water is 1000 kg/m^3 (expected value) at room temperature. If we get a value of 990 kg/m^3 in an experiment, then the error of this experiment is $990 - 1000 = -10 \text{ kg/m}^3$. The relative error is the ratio of the difference divided by the expected value. Therefore, the relative error in this experiment is $|-10|/1000 = 1\%$. The percentage error is the relative error times 100%. The advantage of relative errors is that they have no units.

In statistics, we often use the standard deviation σ_y to characterise the uncertainties or errors of the quantity y. The standard way to write the results of measurements is

$$y = \mu_y \pm \sigma_y, \qquad (13.14)$$

where μ_y is the mean. For example, if we measured the acceleration of gravity at the Earth's surface at a particular location is 9.80m/s^2 with an uncertainty or standard deviation of 0.05 m/s^2, we write

$$g = 9.80 \pm 0.05 \text{ m/s}^2. \qquad (13.15)$$

It is worth pointing out that the form should contain the meaningful significant figures. In this case, we should not write it as $g = 9.800 \pm 0.050$ or 10 ± 0.05.

The standard expression $\mu_y \pm \sigma_y$ is associated with the confidence interval (see Fig. 13.1). This expression suggests that there is about 68.3% of chance that the observations will be in this range. The reason is that the total area under the normal distribution curve is 1, and the area under the curve between $-\sigma$ to $+\sigma$ is

$$A = \int_{-\sigma}^{\sigma} \frac{1}{\sigma\sqrt{2\pi}} e^{-x^2/2\sigma^2} dx = \frac{2}{\sigma\sqrt{2\pi}} \int_{0}^{\sigma} e^{-x^2/2\sigma^2} dx, \qquad (13.16)$$

where we have used the fact that $e^{-x/2\sigma^2}$ is an even function. Using $u = x/\sqrt{2}\sigma$, we have $dx = \sqrt{2}\sigma du$ so that

$$A = \frac{2}{\sigma\sqrt{2\pi}} \int_{0}^{1/\sqrt{2}} \sqrt{2}\sigma e^{-u^2} du = \frac{2}{\sqrt{\pi}} \int_{0}^{1/\sqrt{2}} e^{-u^2} du = \mathrm{erf}(1/\sqrt{2}), \quad (13.17)$$

which is about 0.6827. This means that there is $0.6827 \approx 68.3\%$ chance that the data will fall in the range of $[\mu_y - \sigma_y, \mu_y + \sigma_y]$. In other words, there is about 34.15% chance that it will be in the range from μ_y to $\mu_y + \sigma_y$, and the same chance in the range from $\mu_y - \sigma_y$ to μ_y. Another way of saying this is that the confidence level is 68.3%. If we write the results as $\mu_y + 2\sigma_y$, then the confidence level now becomes 95.4% due that the fact that the area under the curve between -2σ to 2σ is $\mathrm{erf}(\sqrt{2}) \approx 0.954$.

For a given variance σ_x of x for n data points

$$\sigma_x^2 = \frac{1}{n-1} \sum_{i=1}^{n} (x_i - \bar{x})^2, \qquad (13.18)$$

where $\bar{x} = \frac{1}{n} \sum_{i=1}^{n} x_i$ is the sample mean of $x_i(i = 1, 2, ..., n)$, we can calculate the variance of a function $y = f(x)$. We know from the basic definition that

$$\sigma_f^2 = \frac{1}{n-1} \sum_{i=1}^{n} (f_i - \bar{f})^2, \qquad (13.19)$$

where $\bar{f}$ is the mean and $f_i = f(x_i)$. Using $\Delta f = \frac{df}{dx}\Delta x = f'\Delta x$ and setting $\Delta f = f_i - \bar{f}$, we have

$$(f_i - \bar{f}) = (x_i - \bar{x})\frac{df}{dx}. \qquad (13.20)$$

The variance now becomes

$$\sigma_f^2 = \frac{1}{n-1} \sum_{i=1}^{n} \left[(x_i - \bar{x})\frac{df}{dx} \right]^2 = \frac{1}{n-1} \sum_{i=1}^{n} (x_i - \bar{x})^2 \left(\frac{df}{dx}\right)^2. \quad (13.21)$$

In the case of $\frac{df}{dx}$ is independent of x_i or if we evaluate the derivative at $x = \mu_x \approx \bar{x}$, we have

$$\sigma_f^2 = \left(\frac{df}{dx}\right)^2 \frac{1}{n-1} \sum_{i=1}^{n} (x_i - \bar{x})^2 = \sigma_x^2 \left(\frac{df}{dx}\Big|_{\bar{x}}\right)^2, \qquad (13.22)$$

where $|_{\bar{x}}$ means that derivative is evaluated using $x = \bar{x}$. This is the formula for error propagation.

If there are measurements of two independent variables x and y, we can derive similar formulae for the propagation of errors for any function $z = f(x, y)$. From the total derivatives, we know that

$$\Delta f = \frac{\partial f}{\partial x} \Delta x + \frac{\partial f}{\partial y} \Delta y. \tag{13.23}$$

Setting $\Delta f = (f_i - \bar{f})$, $\Delta x = x_i - \bar{x}$ and $\Delta y = y_i - \bar{y}$ where $\bar{y}$ is the mean, we have

$$(f_i - \bar{f}) = (x_i - \bar{x}) \frac{\partial f}{\partial x} + (y_i - \bar{y}) \frac{\partial f}{\partial y}. \tag{13.24}$$

Taking the squares and sums, we have

$$\frac{1}{n-1} \sum_{i=1}^{n} (f_i - \bar{f})^2 = \frac{1}{n-1} \sum_{i=1}^{n} \left[(x_i - \bar{x}) \frac{\partial f}{\partial x} + (y_i - \bar{y}) \frac{\partial f}{\partial y} \right]^2$$

$$= \frac{1}{n-1} \sum_{i=1}^{n} (x_i - \bar{x})^2 (\frac{\partial f}{\partial x})^2$$

$$+ \frac{1}{n-1} \sum_{i=1}^{n} (y_i - \bar{y})^2 (\frac{\partial f}{\partial y})^2 + \frac{2}{n-1} \sum_{i=1}^{n} (x_i - \bar{x})(y_i - \bar{y}) \frac{\partial f}{\partial x} \frac{\partial f}{\partial y}. \tag{13.25}$$

By defining the covariance of x and y as

$$\sigma_{xy} = \frac{1}{n-1} \sum_{i=1}^{n} (x_i - \bar{x})(y_i - \bar{y}), \tag{13.26}$$

and evaluating the derivatives at $x = \bar{x}$ and $y = \bar{y}$, we then have

$$\sigma_f^2 = \frac{1}{n-1} \sum_{i=1}^{n} (f_i - \bar{f})^2 = \sigma_x^2 (\frac{\partial f}{\partial x})^2 + \sigma_y^2 (\frac{\partial f}{\partial y})^2 + 2\sigma_{xy} \frac{\partial f}{\partial x} \frac{\partial f}{\partial y}. \tag{13.27}$$

In the case when x and y are independent so that $\sigma_{xy} = 0$, we have

$$\sigma_f^2 = \sigma_x^2 (\frac{\partial f}{\partial x})^2 + \sigma_y^2 (\frac{\partial f}{\partial y})^2. \tag{13.28}$$

For example, if $z = kx^\alpha y^\beta$ where k, α and β are constants, we have

$$\sigma_z^2 = \sigma_x^2 (k\alpha x^{\alpha-1} y^\beta)^2 + \sigma_y^2 (k\beta x^\alpha y^{\beta-1})^2. \tag{13.29}$$

Dividing both sides by $z^2 = (kx^\alpha y^\beta)^2$ and evaluating the derivatives at $x = \mu_x$, $y = \mu_y$ and $z = \mu_z$, we have the relative errors

$$(\frac{\sigma_z}{\mu_z})^2 = (\frac{\alpha \sigma_x}{\mu_x})^2 + (\frac{\beta \sigma_y}{\mu_y})^2. \tag{13.30}$$

Example 13.3: An experiment gives the following empirical equation

$$z = f(x, y) = x - y,$$

with $\sigma_x = 0.3$ and $\sigma_y = 0.4$. That is $x = 2.5 \pm 0.3$ and $y = 1.0 \pm 0.4$. What is the standard deviation σ_z of $z = f(x, y)$? We know that $\frac{\partial z}{\partial x} = 1$, and $\frac{\partial z}{\partial y} = -1$. From (13.28), we have

$$\sigma_z^2 = \sigma_x^2 \left(\frac{\partial z}{\partial x}\right)^2 + \sigma_y^2 \left(\frac{\partial z}{\partial y}\right)^2 = 0.3^2 \times 1^2 + 0.4^2 \times (-1)^2 = 0.25.$$

Therefore, the standard derivation σ_z is $\sigma_z = \sqrt{0.25} = 0.5$. As the mean $\mu_z = 2.5 - 1.0 = 1.5$, we have $z = 1.5 \pm 0.5$.

As a further example, let us estimate the speed of sound using $v = s/t$. From (13.30), we have $\alpha = 1$ and $\beta = -1$. If the measured distance is $s = 500 \pm 25$ m, and the time taken is 1.50 ± 0.10, we then have

$$\frac{\sigma_v}{\mu_v} = \sqrt{\left(\frac{1 \times 25}{500}\right)^2 + \left(\frac{-1 \times 0.10}{1.5}\right)^2} \approx \sqrt{0.0069} \approx 0.08.$$

As $\mu_v = 500/1.5 \approx 333$, we now get $\sigma_v = \mu_v \times 0.08 \approx 27$. Therefore, we have $v = 333 \pm 27$ m/s.

If we have N observations for the same quantity x each with the same uncertainty σ_x, we have

$$\mu = \frac{x_1 + x_2 + ... + x_N}{N}. \tag{13.31}$$

Since $\frac{\partial \mu}{\partial x_i} = 1/N$, we get

$$\sigma_\mu^2 = \sigma_x^2 \left(\frac{\partial \mu}{\partial x_1}\right)^2 + ... + \sigma_x^2 \left(\frac{\partial \mu}{\partial x_N}\right)^2 = \sigma_x^2 \left(\frac{1}{N}\right)^2 + ... + \sigma_x^2 \left(\frac{1}{N}\right)^2 = \frac{\sigma_x^2}{N}, \tag{13.32}$$

which implies that $\sigma_\mu = \sigma_x/\sqrt{N}$. That is to say, more measurements will reduce the error in the mean, though the standard derivation σ_x remains the same.

13.4   Linear Regression

For experiments and observations, we usually plot one variable such as pressure or price y against another variable x such as time or spatial coordinates. We try to present the data in such a way that we can see some trend in the data. For a set of n data points (x_i, y_i), the usual practice is to try to draw a straight line $y = a + bx$ so that it represents

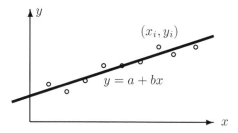

Figure 13.2: Least square and the best fit line.

the major trend. Such a line is often called the regression line or the best-fit line as shown in Figure 13.2.

The method of least squares is to try to determine the two parameters a (intercept) and b (slope) for the regression line from n data points. Assuming that x_i are known more precisely and y_i values obey a normal distribution $p(y_i) \propto \exp\{-[y_i - f(x_i)]^2/2\sigma^2\}$ around the potential best fit line with a variance σ^2, the best fit can be obtained by maximising the likelihood P defined as

$$P = p(y_1) \times p(y_2) \times \ldots \times p(y_n)$$

$$\equiv \prod_{i=1}^{n} p(y_i) = A \exp\left\{ -\frac{1}{2\sigma^2} \sum_{i=1}^{n} [y_i - f(x_i)]^2 \right\}, \qquad (13.33)$$

where A is a constant, and $f(x)$ is the function for the regression [$f(x) = a + bx$ for a simple linear regression].

The essence of the method of least squares is to maximise the probability P by choosing the appropriate a and b. The maximisation of P is equivalent to the minimisation of the exponent ψ

$$\psi = \sum_{i=1}^{n} [y_i - f(x_i)]^2. \qquad (13.34)$$

We see that ψ is the sum of the squares of the deviations $\epsilon_i^2 = (y_i - f(x_i))^2$ where $f(x_i) = a + bx_i$. The minimisation means the least sum of the squares, thus the name of the method of least squares.

In order to minimise ψ as a function of a and b, its derivatives should be zero. That is

$$\frac{\partial \psi}{\partial a} = -2 \sum_{i=1}^{n} [y - (a + bx_i)] = 0, \qquad (13.35)$$

and

$$\frac{\partial \psi}{\partial b} = -2 \sum_{i=1}^{n} x_i [y_i - (a + bx_i)] = 0. \qquad (13.36)$$

By expanding these equations, we have

$$na + b \sum_{i=1}^{n} x_i = \sum_{i=1}^{n} y_i, \tag{13.37}$$

and

$$a \sum_{i=1}^{n} x_i + b \sum_{i=1}^{n} x_i^2 = \sum_{i=1}^{n} x_i y_i, \tag{13.38}$$

which form a system of linear equations for a and b, and it is straightforward to obtain the solutions as

$$a = \frac{1}{n} [\sum_{i=1}^{n} y_i - b \sum_{i=1}^{n} x_i] = \bar{y} - b\bar{x}, \tag{13.39}$$

$$b = \frac{n \sum_{i=1}^{n} x_i y_i - (\sum_{i=1}^{n} x_i)(\sum_{i=1}^{n} y_i)}{n \sum_{i=1}^{n} x_i^2 - (\sum_{i=1}^{n} x_i)^2}, \tag{13.40}$$

where

$$\bar{x} = \frac{1}{n} \sum_{i=1}^{n} x_i, \qquad \bar{y} = \frac{1}{n} \sum_{i=1}^{n} y_i. \tag{13.41}$$

If we use the following notations

$$K_x = \sum_{i=1}^{n} x_i, \quad K_y = \sum_{i=1}^{n} y_i, \quad K_{xx} = \sum_{i=1}^{n} x_i^2, \quad K_{xy} = \sum_{i=1}^{n} x_i y_i, \tag{13.42}$$

then the above equations for a and b become

$$a = \frac{K_{xx} K_y - K_x K_{xy}}{n K_{xx} - (K_x)^2}, \qquad b = \frac{n K_{xy} - K_x K_y}{n K_{xx} - (K_x)^2}. \tag{13.43}$$

The residual error is defined by

$$\epsilon_i = y_i - (a + bx_i), \tag{13.44}$$

whose sample mean is given by

$$\mu_\epsilon = \frac{1}{n} \sum_{i=1}^{n} \epsilon_i = \frac{1}{n} y_i - a - b \frac{1}{n} \sum_{i=1}^{n} x_i = \bar{y} - a - b\bar{x} = 0. \tag{13.45}$$

The sample variance S^2 is

$$S^2 = \frac{1}{n-2} \sum_{i=1}^{n} [y_i - (a + bx_i)]^2, \tag{13.46}$$

where the factor $1/(n-2)$ comes from the fact that two constraints are needed for the best fit, and the residuals therefore have $n-2$ degrees of freedom.

13.5 Correlation Coefficient

The correlation coefficient $r_{x,y}$ is a very useful parameter for finding any potential relationship between two sets of data x_i and y_i for two random variables x and y, respectively. For a set of n data points (x_i, y_i), the correlation coefficient can be calculated by

$$r_{x,y} = \frac{n \sum_{i=1}^{n} x_i y_i - \sum_{i=1}^{n} x_i \sum_{i=1}^{n} y_i}{\sqrt{[n \sum x_i^2 - (\sum_{i=1}^{n} x_i)^2][n \sum_{i=1}^{n} y_i^2 - (\sum_{i=1}^{n} y_i)^2]}}, \quad (13.47)$$

or

$$r_{x,y} = \frac{n K_{xy} - K_x K_y}{\sqrt{(n K_{xx} - K_x^2)(n K_{yy} - K_y^2)}}, \quad (13.48)$$

where $K_{yy} = \sum_{i=1}^{n} y_i^2$.

If the two variables are independent, then $r_{x,y} = 0$, which means that there is no correlation between them. If $r_{x,y}^2 = 1$, then there is a linear relationship between these two variables. $r_{x,y} = 1$ is an increasing linear relationship where the increase of one variable will lead to increase of another. $r_{x,y} = -1$ is a decreasing relationship when one increases while the other decreases.

Example 13.4: Let us try to estimate the b-value in the Gutenberg-Richter law using the data from the Southern California Earthquakes. The numbers of earthquakes from 1987 to 1996 in this region are as follows:

Magnitude above $M =$	3	4	5	6	7
Total numbers	$N = 5454$	561	64	9	1

In order to fit

$$\log_{10}(N) = a + bM,$$

we first have to convert N into $\log_{10} N$, and we have

$y_i = \log_{10} N =$	3.737	2.749	1.806	0.954	0.

Then, we have

$$K_x = \sum_{i=1}^{5} M_i = 3 + 4 + 5 + 6 + 7 = 25, \qquad K_y = \sum_{i=1}^{5} y_i = 9.246,$$

$$K_{xx} = \sum_{i=1}^{5} M_i^2 = 3^2 + 4^2 + 5^2 + 6^2 + 7^2 = 135, \qquad K_{yy} = \sum_{i=1}^{5} y_i^2 = 25.694,$$

and $K_{xy} = \sum_{i=1}^{5} M_i y_i = 36.961$. Therefore, we have

$$a = \frac{K_{xx} K_y - K_x K_{xy}}{n K_{xx} - (K_x)^2} = \frac{135 \times 9.246 - 25 \times 36.961}{5 \times 135 - 25^2} \approx 6.4837,$$

and

$$b = \frac{n K_{xy} - K_x K_y}{n K_{xx} - (K_x)^2} = \frac{5 \times 36.961 - 25 \times 9.246}{5 \times 135 - 25^2} \approx -0.9269.$$

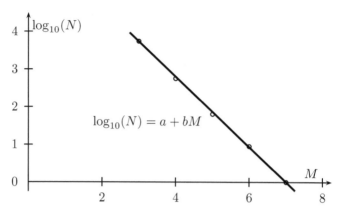

Figure 13.3: Estimating b-value by the best fit to data.

Finally, the correlation coefficient defined by (13.48) becomes

$$r = \frac{5 \times 36.961 - 25 \times 9.246}{\sqrt{(5 \times 135 - 25^2)(5 \times 25.694 - 9.246^2)}} \approx -0.9997.$$

This indeed implies a very strong correlation. The best fit relationship becomes $\log_{10}(N) = 6.4837 - 0.9269M$ (see Fig. 13.3).

There are many important techniques in geostatistics. For example, kriging is a very important tool for analysing the data with spatial correlation and it will provide a good prediction for missing data and interpolation. Interested readers can refer to more advanced literature[1].

13.6   Brownian Motion

Einstein provided in 1905 the first theory of Brownian motion of spherical particles suspended in a liquid, which can be written, after some lengthy calculations, as

$$\overline{\Delta^2} = \frac{k_B}{T} \frac{t}{3\pi\mu a}, \tag{13.49}$$

where a is the radius of the spherical particle, and μ is the viscosity of the liquid. T is the absolute temperature, and t is time. $k_B = R/N_A$ is Boltzmann's constant where R is the universal gas constant and N_A is Avogadro's constant. $\overline{\Delta^2}$ is the mean square of the displacement.

---

[1] For example, Yang, X. S., *Mathematical Modelling for Earth Sciences*, Dunedin Academic, (2008).

Langevin in 1908 presented a very instructive but much simpler version of the theory. If u is the displacement, and $\xi = \frac{du}{dt}$ is the speed at a given instant, then the kinetic energy of the motion should be equal to the average kinetic energy $\frac{1}{2}k_B T$.[2] That is $\frac{1}{2}m\overline{\xi^2} = \frac{1}{2}k_B T$, where m is the mass of the particle. The spherical particle moving at a velocity of ξ will experience a viscous resistance equal to $-6\pi\mu a\xi = -6\pi\mu a\frac{du}{dt}$ according to Stokes' law. If $F(t)$ is the complementary force acting on the particle at the instant so as to maintain the agitation of the particle, we have, according to Newton's second law

$$m\frac{d^2u}{dt^2} = -6\pi\mu a\frac{du}{dt} + F. \tag{13.50}$$

Multiplying both sides by u, we have

$$mu\frac{d^2u}{dt^2} = -6\pi\mu au\frac{d^2u}{dt^2} + Fu. \tag{13.51}$$

Since $(u^2)''/2 = uu'' + u'^2$ or $u\frac{d^2u}{dt^2} = \frac{1}{2}\frac{d^2(u^2)}{dt^2} - (\frac{du}{dt})^2 = \frac{1}{2}\frac{d}{dt}\left[\frac{d(u^2)}{dt}\right] - \xi^2$, we have

$$\frac{m}{2}\frac{d}{dt}\left[\frac{d(u^2)}{dt}\right] - m\xi^2 = -3\pi\mu a\frac{d(u^2)}{dt} + Fu, \tag{13.52}$$

where we have used $uu' = (u^2)'/2$. If we consider a large number of identical particles and take the average, we have

$$\frac{m}{2}\frac{d}{dt}\left[\overline{\frac{d(u^2)}{dt}}\right] - m\overline{\xi^2} = -3\pi\mu a\overline{\frac{d(u^2)}{dt}} + \overline{Fu}, \tag{13.53}$$

where $\overline{()}$ means the average. Since the average of $\overline{Fu} = 0$ due to the fact that the force is random, taking any signs and directions. Let $Z = \overline{d(u^2)/dt}$, we have

$$\frac{m}{2}\frac{dZ}{dt} + 3\pi\mu aZ = k_B T, \tag{13.54}$$

where we have used $m\overline{\xi^2} = k_B T$ discussed earlier. This equation is now known as Langevin's equation. It is a linear first-order differential equation. Its general solution can easily be found using the method discussed in the chapter on differential equations

$$Z_c = Ae^{-t/\tau} + k_B T\frac{1}{3\pi\mu a}, \qquad \tau = \frac{m}{6\pi\mu a}, \tag{13.55}$$

---

[2]Here Langevin considered only one direction. For three directions, the kinetic energy becomes $\frac{3}{2}k_B T$, thus there is a factor 3. This means that equation (13.59) should be $\overline{u^2} = 6Dt$.

where A is the constant to be determined. For the Brownian motion to be observable, t must be reasonably larger than the characteristic time $\tau \approx 10^{-8}$ seconds, which is often the case. So the first term will decrease exponentially and becomes negligible when $t > \tau$ in the standard timescale of our interest. Now the solution becomes

$$Z = \frac{\overline{d(u^2)}}{dt} = k_B T \frac{1}{3\pi\mu a}. \qquad (13.56)$$

Integrating it with respect to t, we have

$$\overline{u^2} = k_B T \frac{t}{3\pi\mu a}, \qquad (13.57)$$

which is Einstein's formula for Brownian motion. Furthermore, the Brownian motion is essentially a diffusion process, and the diffusion coefficient D can be defined as

$$D = \frac{k_B T}{6\pi\mu a}, \qquad (13.58)$$

so that we have

$$\overline{u^2} = 2Dt. \qquad (13.59)$$

This suggests that large particles diffuse more slowly than smaller particles in the same medium. Mathematically, it is very similar to the heat conduction process discussed in earlier chapters.

We know that the diffusion coefficient of sugar in water at room temperature is $D \approx 0.5 \times 10^{-9}$ m^2/s. In 1905, Einstein was the first to estimate the size of sugar molecules using experimental data in his doctoral dissertation on Brownian motion.

Let us now estimate the size of the sugar molecules. Since $k_B = 1.38 \times 10^{-23}$ J/K, $T = 300$ K, and $\mu = 10^{-3}$ Pa s, we have

$$a = \frac{k_B T}{6\pi\mu D} = \frac{1.38 \times 10^{-23} \times 300}{6\pi \times 10^{-3} \times 0.5 \times 10^{-9}} \approx 4.4 \times 10^{-10} \text{ m}, \qquad (13.60)$$

which means that the diameter is about 8.8×10^{-10} m$= 8.8$ nm. In fact, Einstein estimated for the first time the diameter of a sugar molecule was 9.9×10^{-10} m even though the other quantities were not so accurately measured at the time.

From $\overline{u^2} = 2Dt$, we can either estimate the diffusion distance for a given time or estimate the timescale for a given length. For a sugar cube in a cup of water to dissolve completely to form a solution of uniform concentration (without stirring), the time taken will be $t = d^2/2D$, where $d = \sqrt{\overline{u^2}}$ is the size of the cup. Using $d = 10$ cm$= 0.1$ m and $D = 0.5 \times 10^{-9}$ m^2/s, we have $t = 0.1^2/(2 \times 0.5 \times 10^{-9}) \approx 1 \times 10^7$ seconds, which is about 3 months. This is too slow; that is why we always try to stir a cup of tea or coffee.

Chapter 14

Numerical Methods

The beauty of an analytical solution to a problem is that it is accurate and often provides some insight into the main process of interest. However, in most applications, no explicit forms of analytical solutions exist, and we have to use some approximations. In some cases, even the evaluation of a solution is not easy, and we have to use numerical methods to estimate solutions.

There are many different numerical methods for solving a wide range of problems with different orders of accuracy and various levels of complexity. For example, for numerical solutions of ODEs, we can have a simple Euler integration scheme or higher-order Runge-Kutta scheme. For PDEs, we can use finite difference methods, finite element methods, finite volume methods and others. As this book is an introductory textbook, we will only introduce the most basic methods that are extremely useful to a wide range of problems in earth sciences.

To demonstrate how these numerical methods work, we will use step-by-step examples to find the roots of nonlinear equations, to estimate integrals by numerical integration, and to solve ODEs by direct integration and higher-order Runge-Kutta methods.

14.1 Finding the roots

Let us start by trying to find the root of a number a. It is essentially equivalent to finding the solution of

$$x^2 - a = 0. \tag{14.1}$$

We can rearrange it as

$$x = \frac{1}{2}(x + \frac{a}{x}), \tag{14.2}$$

which makes it possible to estimate the root x iteratively. If we start from a guess, say $x_0 = 1$, we can calculate the new estimate x_{n+1} from any previous value x_n using

$$x_{n+1} = \frac{1}{2}(x_n + \frac{a}{x_n}).\tag{14.3}$$

Let us look at an example.

Example 14.1: In order to find $\sqrt{23}$, we have $a = 23$ with an initial guess $x_0 = 1$. The first five iterations are as follows:

$$x_1 = \frac{1}{2}(x_0 + \frac{23}{x_0}) = \frac{1}{2}(1 + \frac{23}{1}) = 12,$$

$$x_2 = \frac{1}{2}(x_1 + \frac{23}{x_1}) = \frac{1}{2}(12 + \frac{23}{12}) \approx 6.9583333.$$

$$x_3 = \frac{1}{2}(x_2 + \frac{23}{x_2}) \approx 5.1318613, \quad x_4 \approx 4.8068329, \quad x_5 \approx 4.79584411.$$

We know that the exact solution is $\sqrt{23} = 4.7958315233$. We can see that after only five iterations, x_5 is accurate to the 4th decimal place.

Iteration is in general very efficient; however, we have to be careful about the proper design of the iteration formula and the selection of an appropriate initial guess. For example, we cannot use $x_0 = 0$ as the initial value. Similarly, if we start $x_0 = 1000$, we have to do more iterations to get the same accuracy.

There are many methods for finding the roots; the Newton-Raphson method is by far the most successful and widely used.

14.2   Newton-Raphson Method

Newton's method is a widely-used classical method for finding the solution to a nonlinear univariate function of $f(x)$ on the interval $[a, b]$. It is also referred to as the Newton-Raphson method. At any given point x_n shown in Fig. 14.1, we can approximate the function by a Taylor series

$$f(x_{n+1}) = f(x_n + \Delta x) \approx f(x_n) + f'(x_n)\Delta x,\tag{14.4}$$

where

$$\Delta x = x_{n+1} - x_n,\tag{14.5}$$

which leads to

$$x_{n+1} - x_n = \Delta x \approx \frac{f(x_{n+1}) - f(x_n)}{f'(x_n)},\tag{14.6}$$

or

$$x_{n+1} \approx x_n + \frac{f(x_{n+1}) - f(x_n)}{f'(x_n)}.\tag{14.7}$$

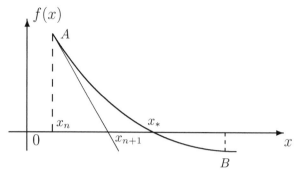

Figure 14.1: Newton's method of approximating the root x_* by x_{n+1} from the previous value x_n.

Since we try to find an approximation to $f(x) = 0$ with $f(x_{n+1})$, we can use the approximation $f(x_{n+1}) \approx 0$ in the above expression. Thus we have the standard Newton iterative formula

$$x_{n+1} = x_n - \frac{f(x_n)}{f'(x_n)}. \tag{14.8}$$

The iteration procedure starts from an initial guess value x_0 and continues until certain criteria are met. A good initial guess will use fewer steps; however, if there is no obvious initial good starting point, you can start at any point on the interval $[a, b]$. But if the initial value is too far from the true zero, the iteration process may fail. So it is a good idea to limit the number of iterations.

Example 14.2:   To find the root of

$$f(x) = x - e^{-\sin x} = 0,$$

we use Newton-Raphson method starting from $x_0 = 0$. We know that

$$f'(x) = 1 + e^{-\sin x} \cos x,$$

and thus the iteration formula becomes

$$x_{n+1} = x_n - \frac{x_n - e^{-\sin x_n}}{1 + e^{-\sin x_n} \cos x_n}.$$

Since $x_0 = 1$, we have

$$x_1 = 0 - \frac{0 - e^{-\sin 0}}{1 + e^{-\sin 0} \cos 0} = 0.5000000000,$$

$$x_2 = 0.5 - \frac{0.5 - e^{-\sin 0.5}}{1 + e^{-\sin 0.5} \cos 0.5} \approx 0.5771952598$$

$$x_3 \approx 0.5787130827, \qquad x_4 \approx 0.5787136435.$$

We can see that x_3 (only three iterations) is very close (to the 6th decimal place) to the true root which is $x_* \approx 0.57871364351972$, while x_4 is accurate to the 10th decimal place.
- -

We have seen that Newton-Raphson's method is very efficient and that is why it is so widely used. Using this method, we can virtually solve almost all root-finding problems. However, this method is not applicable to carrying out integration.

14.3   Numerical Integration

For any smooth function, we can always calculate its derivatives by direct differentiation; however, integration is often difficult even for seemingly simple integrals such as the error function

$$\operatorname{erf}(x) = \frac{2}{\sqrt{\pi}} \int_0^x e^{-u^2} du. \tag{14.9}$$

The integration of this simple integrand $\exp(-u^2)$ does not lead to any simple explicit expression, which is why it is often written as erf(), referred to as the error function. If we pick up a mathematical handbook, we know that $\operatorname{erf}(0) = 0$, and $\operatorname{erf}(\infty) = 1$, while

$$\operatorname{erf}(0.5) \approx 0.52049, \qquad \operatorname{erf}(1) \approx 0.84270. \tag{14.10}$$

If we want to calculate such integrals, numerical integration is the best alternative.

Now if we want to numerically evaluate the following integral

$$\mathcal{I} = \int_a^b f(x)dx, \tag{14.11}$$

where a and b are fixed and finite; we know that the value of the integral is exactly the total area under the curve $y = f(x)$ between a and b. As both the integral and the area can be considered as the sum of the values over many small intervals, the simplest way of evaluating such numerical integration is to divide up the integral interval into n equal small sections and split the area into n thin strips of width h so that $h \equiv \Delta x = (b-a)/n$, $x_0 = a$ and $x_i = ih + a(i = 1, 2, ..., n)$. The values of the functions at the dividing points x_i are denoted as $y_i = f(x_i)$, and the value at the midpoint between x_i and x_{i+1} is labeled as $y_{i+1/2} = f_{i+1/2}$

$$y_{i+1/2} = f(x_{i+1/2}) = f_{i+1/2}, \qquad x_{i+1/2} = \frac{x_i + x_{i+1}}{2}. \tag{14.12}$$

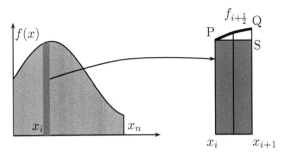

Figure 14.2: Integral as a sum of multiple thin stripes.

The accuracy of such approximations depends on the number n and the way to approximate the curve in each interval.

Figure 14.2 shows such an interval $[x_i, x_{i+1}]$ which is exaggerated in the figure for clarity. The curve segment between P and Q is approximated by a straight line with a slope

$$\frac{\Delta y}{\Delta x} = \frac{f(x_{i+1}) - f(x_i)}{h}, \tag{14.13}$$

which approaches $f'(x_{i+1/2})$ at the midpoint point when $h \to 0$.

The trapezium (formed by P, Q, x_{i+1}, and x_i) is a better approximation than the rectangle (P, S, x_{i+1} and x_i) because the former has an area

$$A_i = \frac{f(x_i) + f(x_{i+1})}{2}h, \tag{14.14}$$

which is closer to the area

$$\mathcal{I}_i = \int_{x_i}^{x_{i+1}} f(x)dx, \tag{14.15}$$

under the curve in the small interval x_i and x_{i+1}. If we use the area A_i to approximate $\mathcal{I}_i$, we have the trapezium rule of numerical integration. Thus, the integral is simply the sum of all these small trapeziums, and we have

$$\mathcal{I} \approx \frac{h}{2}[f_0 + 2(f_1 + f_2 + \dots + f_{n-1}) + f_n]$$

$$= h[f_1 + f_2 + \dots + f_{n-1} + \frac{(f_0 + f_n)}{2}]. \tag{14.16}$$

From the Taylor series, we know that

$$\frac{f(x_i) + f(x_{i+1})}{2} \approx \frac{1}{2}\Big\{[f(x_{i+1/2}) - \frac{h}{2}f'(x_{i+1/2}) + \frac{1}{2!}(\frac{h}{2})^2 f''(x_{i+1/2})]$$

$$+ [f(x_{i+1/2}) + \frac{h}{2}f'(x_{i+1/2}) + \frac{1}{2!}(\frac{h}{2})^2 f''(x_{i+1/2})]\Big\}$$

$$= f(x_{i+1/2}) + \frac{h^2}{8} f''(x_{i+1/2}). \tag{14.17}$$

where $O(h^2 f'')$ means that the value is about the order of $h^2 f''$, or $O(h^2) = Kh^2 f''$ where K is a constant.

The trapezium rule is just one of the simple and popular schemes for numerical integration with the error of $O(h^3 f'')$. If we want higher accuracy, we can either reduce h or use a better approximation for $f(x)$. A small h means a large n, which implies that we have to do the sum of many small sections, and it may increase the computational time.

On the other hand, we can use higher-order approximations for the curve. Instead of using straight lines or linear approximations for curve segments, we can use parabolas or quadratic approximations. For any consecutive three points x_{i-1}, x_i and x_{i+1}, we can construct a parabola in the form

$$f(x_i + t) = f_i + \alpha t + \beta t^2, \qquad t \in [-h, h]. \tag{14.18}$$

As this parabola must go through the three known points (x_{i-1}, f_{i-1}) at $t = -h$, (x_i, f_i) at $t = 0$ and x_{i+1}, f_{i+1} at $t = h$, we have the following equations for α and β

$$f_{i-1} = f_i - \alpha h + \beta h^2, \tag{14.19}$$

and

$$f_{i+1} = f_i + \alpha h + \beta h^2, \tag{14.20}$$

which lead to

$$\alpha = \frac{f_{i+1} - f_{i-1}}{2h}, \qquad \beta = \frac{f_{i-1} - 2f_i + f_{i+1}}{2h^2}. \tag{14.21}$$

In fact, α is the centred approximation for the first derivative f_i' and β is related to the central difference scheme for the second derivative f_i''. Therefore, the integral from x_{i-1} to x_{i+1} can be approximated by

$$\mathcal{I}_i = \int_{x_{i-1}}^{x_{i+1}} f(x)dx \approx \int_{-h}^{h} [f_i + \alpha t + \beta t^2]dt = \frac{h}{3}[f_{i-1} + 4f_i + f_{i+1}],$$

where we have substituted the expressions for α and β. To ensure the whole interval $[a, b]$ can be divided up to form three-point approximations without any point left out, n must be even. Therefore, the estimate of the integral becomes

$$\mathcal{I} \approx \frac{h}{3}[f_0 + 4(f_1 + f_3 + \dots + f_{n-1}) + 2(f_2 + f_4 + \dots + f_{n-2}) + f_n], \tag{14.22}$$

which is the standard Simpson's rule.

As the approximation for the function $f(x)$ is quadratic, an order higher than the linear form, the error estimate of Simpson's rule is thus

$O(h^4)$ or $O(h^4 f''')$ to be more specific. There are many variations of Simpson's rule with higher-order accuracies such as $O(h^5 f^{(4)})$.

Example 14.3: We know the value of the integral

$$I = \text{erf}(1) = \frac{2}{\sqrt{\pi}} \int_0^1 e^{-x^2} dx = 0.8427007929.$$

Let us now estimate

$$I = \frac{2}{\sqrt{\pi}} J, \qquad J = \int_0^1 e^{-x^2} dx,$$

using the Simpson rule with $n = 8$ and $h = (1 - 0)/8 = 0.125$. We have

$$J \approx \frac{h}{3}[f_0 + 4(f_1 + f_3 + f_5 + f_7) + 2(f_2 + f_4 + f_6) + f_8].$$

Since $f_i = e^{-x_i^2} = e^{-(i*h)^2}$, we have $f_0 = 1$, $f_1 = 0.984496$, $f_2 = 0.939413$, $f_3 = 0.868815$, $f_4 = 0.778801$, $f_5 = 0.676634$, $f_6 = 0.569783$, $f_7 = 0.465043$, and $f_8 = 0.367879$. Now the integral estimate of J is

$$J \approx \frac{0.125}{3}[1 + 4 \times 2.9949885 + 2 \times 2.2879967 + 0.367879]$$

$$\approx \frac{0.125}{3} \times 17.923827 \approx 0.746826.$$

Finally, the integral estimate of I is

$$I = \frac{2}{\sqrt{\pi}} J = \frac{2}{\sqrt{3.1415926}} \times 0.746826 \approx 0.842703.$$

We can see that this estimate is accurate to the 5th decimal place.

- -

There are other even better ways for evaluating the integral more accurately using fewer points of evaluation. Such numerical integration is called the Gaussian integration or Gaussian quadrature. Interested readers can refer to more advanced books on numerical methods.

14.4   Numerical Solutions of ODEs

The simplest first-order differential equation can be written as

$$\frac{dy}{dx} = f(x, y), \tag{14.23}$$

where $f(x, y)$ is a known function of x and y. In principle, the solution can be obtained by direct integration

$$y(x) = y(x_0) + \int_{x_0}^x f(x, y(x))dx, \tag{14.24}$$

but in practice it is usually impossible to do the integration analytically, as it requires the solution $y(x)$ to evaluate the right-hand side. Numerical integration in this case is the most common technique for obtaining approximate solutions. A naive approach is to use the standard techniques such as Simpson's rule for numerical integration to evaluate the integral numerically. However, since we have to use some approximation for $y(x)$, such techniques hardly work here without modification. A better approach is to start from the known initial value $y(x_0)$ at x_0, and try to march to the next point at $x = x_0 + h$ where h is a small increment. In this manner, the solution at all other values of x can be estimated numerically. This is essentially the Euler scheme.

14.4.1　Euler Scheme

The basic idea of the Euler method is to approximate the integral (14.24) using an iterative procedure

$$y_{n+1} = y_n + \int_{x_n}^{x_{n+1}} f(x, y)dx = y_n + hf(x_n, y_n), \qquad (14.25)$$

where $h = \Delta x = x_{n+1} - x_n$ is a small increment. Here we use the notations $x_n = x_0 + nh (n = 0, 1, 2, ..., N)$ and $y_n = y(x_n)$. This is essentially to divide the interval from x_0 to $x = Nh$ into N small interval of width h, so the value of $f(x, y)$ is approximated by its value in the interval as $f(x_n, y_n)$.

The above equation can also be viewed as the approximation of the first derivative

$$\frac{dy_n}{dx} = y'_n = \frac{y_{n+1} - y_n}{h}. \qquad (14.26)$$

This is a forward difference scheme for approximating the first derivative. From the differentiation and integration discussed earlier in this book, we know that such an approximation is very crude, and the accuracy is at most $O(h^2)$. A potentially better way is to use the so-called central difference to approximate the first derivative

$$y'_n = \frac{y_{n+1} - y_{n-1}}{2h}, \qquad (14.27)$$

which uses two steps. This does not always solve the problem. In order to get a reasonable accuracy, we have to use a very small value of h. There is also an issue called numerical stability. If the step size h is too large, there is a possibility that any error will be amplified during each iteration, and after some iterations, the value y_n might be unstable and become meaningless. A far better and more stable numerical scheme is the well-known Runge-Kutta method.

14.4.2   Runge-Kutta Method

The Runge-Kutta method uses a trial step to march the solution to the midpoint of the interval by the central difference

$$y_{n+1/2} = y_n + \frac{h}{2} f(x_n, y_n), \tag{14.28}$$

then it combines with the standard forward difference as used in Euler scheme

$$y_{n+1} = y_n + h f(x_{n+1/2}, y_{n+1/2}). \tag{14.29}$$

This scheme can be viewed as a predictor-corrector method as it first predicts a value in the midpoint, and then corrects it to an even better estimate at the next step.

   If we follow the same idea, we can devise higher-order methods with multiple steps, which will give higher accuracy. The popular classical Runge-Kutta method is a fourth-order method, which involves four steps to advance from n to $n + 1$. We have

$$a = h f(x_n, y_n), \qquad b = h f(x_n + h/2, y_n + a/2),$$

$$c = h f(x_n + h/2, y_n + b/2), \qquad d = h f(x_n + h, y_n + c),$$

$$y_{n+1} = y_n + \frac{a + 2(b + c) + d}{6}, \tag{14.30}$$

which is fourth-order accurate.

Example 14.4:   Let us solve the nonlinear equation numerically

$$\frac{dy}{dx} + y^2 = -1, \qquad x \in [0, 2]$$

with the initial condition $y(0) = 1$. We know that $\frac{dy}{dx} = -(1 + y^2)$, or

$$-\int \frac{1}{1 + y^2} dy = -\tan^{-1} y = \int dx = x + K,$$

where K is the constant of the integration. This gives

$$y = -\tan(x + K).$$

Using the initial condition $y = 1$ and $x = 0$, we have $1 = -\tan(K)$, or

$$K = -\tan^{-1}(1) = -\frac{\pi}{4}.$$

The analytical solution becomes

$$y(x) = -\tan(x - \frac{\pi}{4}).$$

On the interval $[0, 2]$, let us first solve the equation using the Euler scheme for $h = 0.5$. There are five points $x_i = ih$ (where $i = 0, 1, 2, 3, 4$). As $dy/dx = f(y) = -1 - y^2$, we have the Euler scheme

$$y_{n+1} = y_n + hf(y_n) = y_n - h - hy_n^2.$$

From the initial condition $y_0 = 1$, we now have

$$y_1 = y_0 - h - hy_0^2 = 1 - 0.5 - 0.5 \times 1^2 = 0,$$

$$y_2 \approx -0.5, \qquad y_3 \approx -1.125, \qquad y_4 = -2.2578.$$

These are significantly different (about 30%) from the exact solution

$$y_0* = 1, \quad y_1^* = 0.2934079, \quad y_2^* = -0.21795809,$$

$$y_3^* = -0.86756212, \quad y_4^* = -2.68770693.$$

Now let us use the Runge-Kutta method to solve the same equation to see if it is better. Since $f(x_n, y_n) = -1 - y_n^2$, we have

$$a = hf(x_n, y_n) = -h(1 + y_n^2), \qquad b = -h[1 + (y_n + \frac{a}{2})^2],$$

$$c = -h[1 + (y_n + \frac{b}{2})^2], \qquad d = -h[1 + (y_n + c)^2],$$

and

$$y_{n+1} = y_n + \frac{a + 2(b + c) + d}{6}.$$

From $y_0 = 1$ and $h = 0.5$, we have

$$y_1 \approx 0.29043, \ y_2 \approx -0.22062, \ y_3 = -0.87185, \ y_4 \approx -2.67667.$$

These values are within about 1% of the analytical solution y_n^*. We can see that even with the same step size h, the Runge-Kutta method is more efficient and accurate than the Euler scheme.

- -

Generally speaking, higher-order schemes are better than lower-order schemes, but not always. Let us look at an example.

Example 14.5: We know from the example in an earlier chapter on ODEs that the solution to the differential equation

$$\frac{dy}{dx} = (x^4 + e^x)e^{-y},$$

with the initial condition $y(0) = 0$ at $x_0 = 0$ is

$$y = \ln(e^x + \frac{1}{5}x^5).$$

Now let us try to solve it numerically to estimate $y(1)$.

First, we divide the interval $[0, 1]$ into two steps so that $h = (1-0)/2 = 0.5$. We have $x_0 = 0$ and $y_0 = 0$. Then, we use the Euler scheme, and we have y_1 at $x_1 = 0.5$

$$y_1 = y_0 + hf(x_0, y_0) = y_0 + h(x_0^4 + e^{x_0})e^{-y_0} = 0 + 0.5 \times (0^4 + e^0)e^{-0} = 0.5.$$

The next step is to evaluate y_2 at $x_2 = 2h = 1$, and we have

$$y_2 = y_1 + h(x_1^4 + e^{x_1})e^{-y_1} = 0.5 + 0.5 \times (0.5^4 + e^{0.5})e^{-0.5} \approx 1.0189.$$

We know the exact value at $x = 1$ is

$$y_* = y(1) = \ln(e^1 + \frac{1^5}{5}) = 1.07099503.$$

We can see that the value y_2 obtained by the Euler scheme is about 5% from the true solution. Now let us see whether the Runge-Kutta method can do better.

For the first step to get y_1, we have

$$a = hf(x_0, y_0) = h(x_0^4 + e^{x_0})e^{-y_0} = 0.5 \times (0^4 + e^0) \times e^{-0} = 0.5,$$

$$b = hf(x_0 + h/2, y_0 + a/2) = 0.5 \times [(0+h/2)^4 + e^{0+h/2}] \times e^{-(0+a/2)} = 0.50152,$$

and

$$c = hf(x_0 + h/2, y_0 + b/2) \approx 0.50114, \quad d = hf(x_0 + h, y_0 + c) \approx 0.518363.$$

Now we get

$$y_1 = y_0 + \frac{a + 2(b+c) + d}{6} = 0.503947.$$

The next step to get y_2 from $x_1 = 0.5$ and $y_1 = 0.503947$, we have

$$a = hf(x_1, y_1) = 0.5 \times (0.5^4 + e^{0.5}) \times e^{-0.503947} \approx 0.51691,$$

$$b = hf(x_1 + \frac{h}{2}, y_1 + \frac{a}{2}) \approx 0.56765, \quad c = hf(x_1 + \frac{h}{2}, y_1 + \frac{b}{2}) \approx 0.55343,$$

and $d = hf(x_1 + h, y_1 + c) \approx 0.64580$. Finally, we have

$$y_2 = y_1 + \frac{a + 2(b+c) + d}{6} \approx 1.07142.$$

We can see that Runge-Kutta is more accurate than the Euler method, and the solution is about 0.04% from the true value. However, this Runge-Kutta method is of $O(h^4)$ accurate.

- -

This example demonstrates that higher-order methods are usually the best choice, and they generally work better for most problems.

That is why they are more widely used and have been implemented in many software packages. The general guidance is that you should try to use these methods first until you find a better method for a particular problem.

So far, we have only introduced the numerical method to the first-order equations. Higher-order ordinary differential equations can always be converted to a first-order system. For example, the following second-order equation

$$y''(x) + p(x)y'(x) + q(x)y(x) = r(x), \qquad (14.31)$$

can be rewritten as a system if we let $u = y'(x)$

$$y' = u, \qquad u' = r(x) - q(x)y(x) - p(x)u. \qquad (14.32)$$

The above system of two equations can be written as

$$\frac{d}{dx}\begin{pmatrix} y \\ u \end{pmatrix} = \begin{pmatrix} u \\ r(x) - q(x)y - p(x)u \end{pmatrix}. \qquad (14.33)$$

This is essentially the same as (14.23) in the context of vectors and matrices, and it can essentially be solved using the similar method with little modifications.

Numerical solutions of partial differential equations are even more difficult, especially when the domain is multidimensional and irregular. There are various methods for solving PDEs, including finite different methods, finite volume methods, finite element methods, boundary element methods, spectral methods, and meshless methods.

The literature about numerical methods is vast, especially about the solution of partial differential equations with complex boundary conditions and irregular geometry. Interested readers can refer to more advanced textbooks. For a good start, the book *Mathematical Modelling for Earth Sciences* by the same author is recommended as it covers all the main numerical methods and more advanced mathematical tools in earth sciences.

Bibliography

References

[1] Amundson J. M., Truffer M., and Luthi M. P., Time-dependent basal stress conditions beneath Black Rapids Glacier, Alaska, USA, inferred from measurements of ice deformation and surface motion, *J. of Glaciology*, **52**, 347-357 (2006).

[2] Beard K. V., Terminal velocity and shape of cloud and precipitation drops aloft, *J. Atmosp. Sci.*, **33**, 851-864 (1976).

[3] Bercovic D., Ricard Y., and Richards M. A., The relation between mantle dynamics and plate tectonics: A primer, in: *The History and Dynamics of Global Plate Motions* (Eds Richards M. A., Gordon R. G., van de Hilst R. O.), AGU Geophysical Monograph 121, AGU Publications (2000).

[4] Berger A. L., Long term variations of the Earth's orbital elements, *Celestial Mechanics*, **15**, 53-74 (1977).

[5] Brown P., Spalding R. E., ReVelle D. O., Tagliaferri E., Worden S. P., The flux of small near-Earth objects colliding with the Earth, *Nature*, **420**, 294-296 (2002).

[6] Committee on Grand Research Questions in the Solid-Earth Sciences, National Research Council, *Origin and Evolution of Earth: Research Questions for a Changing Planet*, National Academies Press, (2008).

[7] Doin M. P. and Fleitout, Thermal evolution of the oceanic lithosphere: An alternate view, *Earth Planet. Sci. Lett.*, **142**, 121-136 (1996).

[8] Drew D. A., Mathematical modelling of two-phase flow, *A. Rev. Fluid Mech.*, **15**, 261-291 (1983).

[9] England P., Molnar P., and Richter F., John Perry's neglected critique of Kelvin's age for the Earth: A missed opportunity in geodynamics, *GSA Today*, **17**, 4-9 (2007).

[10] Fowler A. C., A mathematical model of magma transport in the asthenosphere, *Geophys. Astrophys. Fluid Dyn.*, **33**, 63-96 (1985).

[11] Fowler A. C. and Yang X. S., Fast and slow compaction in sedimentary basins, *SIAM J. Appl. Math.*, **59**, 365-385 (1998).

[12] Fowler A. C. and Yang X. S., Pressure solution and viscous compaction in sedimentary basins, *J. Geophys. Res.*, **104**, 12989-12997 (1999).

[13] Fowler A. C. and Yang X. S., Dissolution-precipitation mechanism for diagenesis in sedimentary basins, *J. Geophys. Res.*, **108**, 10.1029/2002JB002269 (2003).

[14] Holter O., Ingebretsen F., and Kanestrom I., Analytic CO_2 model calculations and global temperature, *Eur. J. Phys.*, **20**, 483-494 (1999).

[15] Krige D. G., A statistical approach to some basic mine valuation problems on the Witwatersrand, *J. of the Chem., Metal. and Mining Soc. of South Africa*, **52**, 119-139 (1951).

[16] Lemons D. S., Gythiel A., Paul Langevin's 1908 paper 'On the Theory of Brownian Motion', *Am. J. Phys.*, **65**, 1079-1081 (1997).

[17] Mason B. J., Physics of a raindrop, *Phys. Educ.*, **13**, 414-419 (1978).

[18] Melosh H. J., *Impact Cratering*, Oxford University Press, (1989).

[19] Milankovitch M., Canon of isolation and the ice-age problem, *K. Serb. Adad. Geogr.*, special publication, **132**, 484 (1941).

[20] NASA's Near-Earth Object Science Definition Team, *Study to Determine the Feasibility of Extending the Search for Near-Earth Objects to Smaller Limiting Diameters*, (2003).

[21] Parsons B. and Sclater J. G., Analysis of the variation of ocean floor bathymetry and heat flow with age, *J. Geophys. Res.*, **82**, 803-827(1977).

[22] Peltier W. R., Mantle viscosity and ice-age ice sheet topography, *Science*, **273**, 1359-1364 (1996).

[23] Price, R., The mechanical paradox of large overthrusts, GSA Bull., **100**, 1989-1908 (1988).

[24] Revil A., Pervasive pressure-solution transfer: a poro-visco-plastic model, *Geophys. Res. Lett.*, **26**, 255-258 (1999).

[25] Richter F., Dynamical models for sea floor spreading, *Rev. Geophys. Space Phys.*, **11**, 223-287 (1973).

[26] Southern California Earthquake Data Centre, www.data.scec.org/

[27] Turcotte D. L. and Schubert G., *Geodynamics: Applications of Continuum Physics to Geological Problems*, Wiley & Sons, New York, (1982).

[28] USGS Hawaiian Volcano Observatory, hvo.wr.usgs.gov/volcanoes/

[29] Waltham D., *Mathematics: a Simple Tool for Geologists*, Routeldge (1997).

[30] Weisstein E. W., http://mathworld.wolfram.com

[31] Wikipedia, http://en.wikipedia.com

[32] Yang X. S. and Young Y., Cellular automata, PDEs and pattern formation (Chapter 18), in *Handbook of Bioinspired Algorithms*, edited by Olarius S. and Zomaya A., Chapman & Hall/CRC, (2005).

Further Reading

[33] Abramowitz M. and Stegun I. A., *Handbook of Mathematical Functions*, Dover Publication, (1965).

[34] Adamatzky A. and Teuscher C., *From Utopian to Genuine Unconventional Computers*, Luniver Press, (2006).

[35] Arfken G., *Mathematical Methods for Physicists*, Academic Press, (1985).

[36] Armstrong M., *Basic Linear Geostatistics*, Springer (1998).

[37] Atluri S. N., *Methods of Computer Modeling in Engineering and the Sciences*, Vol. I, Tech Science Press, (2005).

[38] Carrrier G. F. and Pearson C. E., *Partial Differential Equations: Theory and Technique*, 2nd Edition, Academic Press, (1988).

[39] Carslaw H. S. and Jaeger J. C., *Conduction of Heat in Solids*, 2nd Ed, Oxford University Press, (1986).

[40] Courant R. and Hilbert, D., *Methods of Mathematical Physics*, 2 volumes, Wiley-Interscience, New York, (1962).

[41] Crank J., *Mathematics of Diffusion*, Clarendon Press, Oxford, (1970).

[42] Cook R. D., *Finite Element Modelling For Stress Analysis*, Wiley & Sons, (1995).

[43] Devaney R. L., *An Introduction to Chaotic Dynamical Systems*, Redwood, (1989).

[44] Fenner R. T., *Engineering Elasticity*, Ellis Horwood Ltd, (1986).

[45] Farlow S. J., *Partial Differential Equations for Scientists and Engineers*, Dover Publications, (1993).

[46] Fletcher, C. A. J. and Fletcher C. A., *Computational Techniques for Fluid Dynamics*, Vol. I, Springer-Verlag, GmbH, (1997).

[47] Forsyth A. R., *Calculus of Variations*, New York, Dover (1960).

[48] Fowler A. C., *Mathematical Models in the Applied Sciences*, Cambridge University Press, (1997).

[49] Gardiner C. W., *Handbook of Stochastic Methods*, Springer, (2004).

[50] Gershenfeld N., *The Nature of Mathematical Modeling*, Cambridge University Press, (1998).

[51] Gill P. E., Murray W., and Wright M. H., *Practical optimisation*, Academic Press Inc, (1981).

[52] Gleick J., *Chaos: Making a New Science*, Penguin, (1988).

[53] Goodman R., *Teach Yourself Statistics*, London, (1957).

[54] Greenkorn R. A., *Flow phenomena in porous media*, Marcel Dekker Inc, (1983).

[55] Happel J. and Brenner H., *Low Reynolds Number Hydrodynamics*, Nijhoff, Dordrecht, (1983).

[56] Hinch E. J., *Perturbation Methods*, Cambridge Univ. Press, (1991).

[57] Jeffrey A., *Advanced Engineering Mathematics*, Academic Press, (2002).

[58] John F., *Partial Differential Equations*, Springer, New York, (1971).

[59] Keary P. A., and Vine F. J., *Global Tectonics*, 2nd Eds, Blackwell Publishing, (1996).

[60] Keener J. and Sneyd J., *A Mathematical Physiology*, Springer-Verlag, New York, (2001).

[61] Kitanidis P. K., *Introduction to Geostatistics*, Cambridge University Press, (1997).

[62] Korn G. A. and Korn T. M., *Mathematical Handbook for Scientists and Engineers*, Dover Publication, (1961).

[63] Kreyszig E., *Advanced Engineering Mathematics*, 6th Edition, Wiley & Sons, New York, (1988).

[64] Langtangen H. P., *Computational Partial Differential Equations: Numerical Methods and Diffpack Programming*, Springer, (1999).

[65] LeVeque R. J., *Finite Volume Methods for Hyperbolic Problems*, Cambridge University Press, (2002).

[66] Mitchell A. R. and Griffiths D. F., *Finite Difference Method in Partial Differential Equations*, Wiley & Sons, New York, (1980).

[67] Moler C. B., *Numerical Computing with MATLAB*, SIAM, (2004).

[68] Murch B. W. and Skinner B. J., *Geology Today - Understanding Our Planet*, John Wiley & Sons, (2001).

[69] Ockendon J., Howison S., Lacey A. and Movchan A., *Applied Partial Differential Equations*, Oxford University Press, (2003).

[70] Pallour J. D. and Meadows D. S., *Complex Variables for Scientists and Engineers*, Macmillan Publishing Co., (1990).

[71] Papoulis A., *Probability and statistics*, Englewood Cliffs, (1990).

[72] Pearson C. E., *Handbook of Applied Mathematics*, 2nd Ed, Van Nostrand Reinhold, New York, (1983).

[73] Petrenko V. F. and Whitworth R. W., *Physics of Ice*, Oxford University Press, (1999).

[74] Press W. H., Teukolsky S. A., Vetterling W. T. and Flannery B. P., *Numerical Recipes in C++: The Art of Scientific Computing*, 2nd Ed., Cambridge University Press, (2002).

[75] Puckett E. G. and Colella P., *Finite Difference Methods for Computational Fluid Dynamics*, Cambridge University Press, (2005).

[76] Ramsay J. G. and Huber M. I., *The Techniques of Modern Structural Geology*, vol. 2, Academic Press (1987).

[77] Riley K. F., Hobson M. P. and Bence S. J., *Mathematical Methods for Physics and Engineering*, 3rd Ed., Cambridge University Press (2006).

[78] Ross S., *A First Course in Probability*, 5th Ed., Prentice-Hall, (1998).

[79] Ryan M. P., *Magma Transport and Storage*, Wiley & Sons, (1991).

[80] Selby S. M., *Standard Mathematical Tables*, CRC Press, (1974).

[81] Shi G. R., *Numerical Methods for Petroliferous Basin Modeling*, Petroleum Industry Press, Beijing (2000).

[82] Smith G. D., *Numerical Solutions of Partial Differential Equations: Finite Differerence Methods*, 3rd ed., Clarendon Press, Oxford, (1985).

[83] Turcotte D. L. and Schubert G., *Geodynamics: Applications of Continuum Physics to Geological Problems*, Wiley & Sons, New York, (1982).

[84] Wang H. F., *Theory of Linear Poroelasticity: with applications to geomechanics and hydrogeology*, Princeton University Press, (2000).

[85] Wylie C. R., *Advanced Engineering Mathematics*, Tokyo, (1972).

[86] Versteeg H. K. and Malalasekra W., *An Introduction to Computational Fluid Dynamics: The Finite Volume Method*, Prentice Hall, (1995).

[87] Yang X. S., *Theoretical Basin Modelling*, Exposure Publishing, (2006).

[88] Yang X. S., *Mathematical Modelling for Earth Sciences*, Dunedin Academic, (2008).

[89] Yang X. S., *Introduction to Computational Mathematics*, World Scientific, (2008).

[90] Zienkiewicz O. C. and Taylor R. L., *The Finite Element Method*, vol. I/II, McGraw-Hill, 4th Edition, (1991).

Appendix A

Glossary

Absolute value: The value of a real number disregarding its sign, thus it is always non-negative. For example, $|+2| = 2$, $|-5| = 5$, and $|0| = 0$.

Algorithm: A step-by-step description of a solution procedure to a problem, though it means numerical algorithms in computing.

Auxiliary equation: An algebraic equation, also called **characteristic equation**, derived from a homogeneous ordinary differential equation. For example, from the differential equation $ay'' + by' + cy = 0$ where a, b, c are constants, we use $y = \exp(\lambda x)$, and we have its corresponding auxiliary equation $a\lambda^2 + b\lambda + c = 0$.

Binomial theorem: The expansion of $(a+b)^n$ for positive integer n. Its coefficients are commonly expressed in terms of $\binom{n}{k} = \frac{n!}{k!(n-k)!}$.

Chain rule: A rule for differentiation. If a complex function $f(x)$ can be expressed as $f[u(x)]$, the chain rule states $\frac{df}{dx} = \frac{df}{du} \cdot \frac{du}{dx}$. For example, $[\exp(-x^2)]' = \exp(-x^2) \times (-2x) = -2x\exp(-x^2)$.

Characteristic equation: A equation associated with eigenvalues of a matrix $\boldsymbol{A}$. The characteristic equation is written as $\det(\boldsymbol{A} - \lambda \boldsymbol{I}) = |\boldsymbol{A} - \lambda \boldsymbol{I}| = 0$. It also means the auxiliary equation for ordinary differential equations.

Complementary function: A function or solution to a homogeneous ordinary differential equation. For example, the homogeneous equation of $y''(x) + 5y'(x) - 6y(x) = \sin x$ is $y''(x) + 5y'(x) - 6y(x) = 0$ whose complementary function is $y_c = Ae^{2x} + Be^{-3x}$ where A and B are undetermined coefficients.

Complex conjugate: For any complex number $z = a + bi$, its complex conjugate is $z^* = a - bi$. For example, for $z = 2 - 5i$, we have $z^* = 2 + 5i$.

Complex numbers: A pair of numbers which can be written as $a + ib$ where a and b are real numbers, and $i = \sqrt{-1}$ or $i^2 = -1$. Here

a is called the real part, while b is called the imaginary part. So both $-2 - 3i$ and $5i$ are complex numbers.

Cosine rule: For a triangle with two sides b and c as well as an angle A between them, the third side can be calculated by the cosine rule $a^2 = b^2 + c^2 - 2bc \cos A$.

de Moivre's formula: For any real number θ and integer n, De Moivre's formula is $(\cos\theta + i\sin\theta)^n = \cos(n\theta) + i\sin(n\theta)$.

Descartes' theorem: A theorem relates the number of roots of a polynomial with the number of sign changes in its coefficients.

Derivative: The gradient of a function or another derivative. For example, df/dx is the gradient of $f(x)$, while the second derivative d^2f/dx^2 is the gradient of the first derivative df/dx.

Determinant: A quantity associated with a square matrix. For example, $\begin{vmatrix} a & b \\ c & d \end{vmatrix} = ad - bc$, $\begin{vmatrix} 3 & 1 \\ 4 & -2 \end{vmatrix} = 3 \times (-2) - 1 \times 4 = -10$.

Differential equation: A mathematical equation that contains derivatives. It includes both ordinary differential equation and partial differential equations.

Displacement: A vector pointing from the initial location to the final location.

Dot product: The dot product is also called the inner product. For any two vectors $\boldsymbol{u} = (a \quad b)^T$ and $\boldsymbol{v} = (c \quad d)^T$, their dot product is $\boldsymbol{u} \cdot \boldsymbol{v} = ac + bd$.

Equation: A mathematical statement using an equal '=' sign. For example, $x^2 - 1 = 5$ and $a^2 + b^2 = c^2$ are equations.

Error function: A function defined by $\mathrm{erf}(x) = \frac{2}{\sqrt{\pi}} \int_0^x e^{-u^2} du$. This means that $\mathrm{erf}(0) = 0$, $\mathrm{erf}(\infty) = 1$ and $\mathrm{erf}(-\infty) = -1$.

Exponential function: A function in the form a^x where x is the exponent and a is the base. A very widely used base is $e \approx 2.71828$ and e^x is often written as $\exp(x)$.

Euler's formula: The formula for complex numbers: $e^{ix} = \cos x + i\sin x$. Here x is in radians.

Factorial: A factorial is a product defined by $n! = 1 \times 2 \times ... \times n$. For example, $4! = 1 \times 2 \times 3 \times 4 = 24$.

Fraction: A ratio of two integers a, b in the form a/b or $\frac{a}{b}$ where the denominator b cannot be zero. For example, $-2/3$, and $22/7$.

Fourier series: Any piecewise function $f(t)$ in $[-T, T]$ can be expressed as a Fourier series $f(t) = \frac{a_0}{2} + \sum_{n=1}^{\infty}[a_n \cos(n\pi t/T) + b_n \sin(n\pi t/T)]$ where a_n, b_n are the Fourier coefficients.

Function: A mathematical relationship between an independent variable such as x and a dependent or response variable such as y, often written as $y = f(x)$. For example, $y = x^2 - 2x$ is a function.

Gaussian distribution: A very widely used probability distribution, also called the normal distribution $p(x) = \frac{1}{\sigma\sqrt{2\pi}}\exp[-\frac{(x-\mu)^2}{2\sigma^2}]$ with a mean of μ and a standard deviation of σ.

Gradient: The rate of change of a curve or a function $f(x)$, often written as the first derivative df/dx.

Hyperbolic functions: Functions such as $\sinh(x)$, $\cosh(x)$ and $\tanh(x)$.

Ideal gas law: A law for ideal gas, often expressed as $pV = nRT$ where p, V, T are the pressure, volume and temperature of the gas, respectively. n is the number of moles and $R = 8.31$ J/K mol is the universal gas constant.

Identity: An equality that is true for all values of the variables concerned, often use the notation $\equiv$ in stead of $=$. For example, $\sin^2\theta + \cos^2\theta \equiv 1$ is true for all θ.

Identity matrix: A square matrix with all diagonal elements being 1s and off diagonal elements being 0s. For example, a 2×2 identity matrix is $I = \begin{pmatrix} 1 & 0 \\ 0 & 1 \end{pmatrix}$.

Index form: A concise form to write repeated multiplications such as $a^5 = a \times a \times a \times a \times a$.

Integer: whole numbers such as $-2, -1, 0, +1, +5$, and $+56789$.

Integral: The reverse of differentiation, often written as $\int f(x)dx$ where $f(x)$ is called the integrand. For example, $\int x^n dx$ is an indefinite integral, while $\int_a^b \sin(x)dx$ is a definite integral with a and b being its integration limits.

Integration by parts: A technique for integration using $\int u dv = uv - \int v du$.

Inverse: An inverse of a function $y = f(x)$ is to find $x = g(y)$. For example, $y = \sin x$, we have $x = \sin^{-1} y$. For a square matrix A, its inverse is A^{-1} if A is not singular or $\det(A) \neq 0$.

Iteration: A procedure to obtain a solution numerically.

Irrational numbers: Numbers such as $\sqrt{2}$ and π that cannot be expressed as a fraction.

Limit: A concept to describe the behaviour of a function when its argument or independent variable gets close to a point of interest. For example, $\lim_{x\to 0} \sin(x)/x = 1$. Limits can also mean the bounds of an interval.

Linear regression: A method to best-fit a set of data (x_n, y_n) to a linear equation $y = a + bx$.

Linear system: A set of multiple linear equations. For example, $x - 2y = 1$ and $x + 2y = 5$ form a simple linear system which can be written as the matrix form $\begin{pmatrix} 1 & -2 \\ 1 & 2 \end{pmatrix}\begin{pmatrix} x \\ y \end{pmatrix} = \begin{pmatrix} 1 \\ 5 \end{pmatrix}$.

Logarithm: A mathematical function: $y = \log_b(x)$ where b is the base. Commonly used bases are $b = 10$, and $b = e \approx 2.71828$. In the latter case, the logarithm is called the natural logarithm, and written as $y = \ln(x)$.

Matrix: An rectangular array of numbers. For example, $\begin{pmatrix} 2 & -3 \\ -5 & 0 \end{pmatrix}$ is a 2×2 matrix. A vector is a special case of a matrix.

Mean: A statistical term which is the average of all values. For example, the mean of $1, 2, 5, -6$ is $(1 + 2 + 5 + (-6))/4 = 0.5$.

Modulus: For a real number, its modulus is its absolute value. For a complex number $a + bi$, its modulus is the length or $|a + bi| = \sqrt{a^2 + b^2}$.

Natural numbers: Positive integers $1, 2, 3, ...$, or non-negative integers $(0, 1, 2, ...)$.

Newton-Raphson method: A numerical method for solving or finding the root(s) of a nonlinear function.

Normal distribution: A probability distribution. See **Gaussian distribution**.

Numerical integration: A numerical method to approximate or evaluate the value of a complex integral.

ODE: An ordinary differential equation.

Ordinary differential equations: A relationship or equation including derivatives. For example, $dy/dx = -\exp(-x)$ is an ordinary differential equation.

Partial derivative: For a function of two or more independent variables, its partial derivative is the derivative with respect to one of those variables. For example, $f(x, y) = xy + y^2$, we have $\frac{\partial f}{\partial x} = y$ and $\frac{\partial f}{\partial y} = x + 2y$.

Particular integral: Any function that satisfies an ordinary differential equation. For example, the particular integral of $y''(x) + 5y'(x) - 6y(x) = x - 2$ is $y_p = -x/6 + 7/36$.

PDE: A partial differential equation.

Partial differential equation: A relationship or equation which contains partial derivatives such as $\frac{\partial u}{\partial x}$ and/or derivatives. For example, $\frac{\partial u}{\partial t} = D\frac{\partial^2 u}{\partial x^2}$ is a partial differential equation.

Poisson's distribution: A probability distribution for discrete events. $p(k) = \lambda^k e^{-\lambda}/k!$ where $k = 0, 1, 2, ...$ and $\lambda > 0$.

Poisson's ratio: A material property ν which is the ratio of the transverse strain to the axial strain. Most rocks have a Poisson's ratio in the range of 0.2 to 0.4. For example, $\nu \approx 0.3$ for shales, 0.2 for sandstones and 0.4 for coal.

Polar form: A way of expressing a complex number $z = a + bi$ using $z = re^{i\theta}$ where r is the modulus of z and θ is the angle made with the real axis.

Polynomial: An expression with a finite number of terms involving the powers of x. For example, $x^2 - 2x + 3$ a quadratic polynomial while $x^5 - x^3 - x$ is a polynomial of degree 5.

Power function: A mathematical function in the form x^a where the exponent a is a real number and x is the independent variable.

Probability: A number or expected frequency of an even occurring. It is always in the range of $[0, 1]$.

Pythagoras' theorem: For a right-angled triangle, the length of the hypotenuse c is related to the lengths of the other two sides by Pythagoras' theorem: $a^2 + b^2 = c^2$.

Radian: A unit of angle which is equal to $180°/\pi \approx 57.2958°$. So a right angle $90°$ is $\pi/2$ in radians. Conversely, $1° = \pi/180 \approx 0.01745$ radians.

Random variable: A variable to represent the outcome of an event such as the noise level or the number of earthquakes in a period.

Rational numbers: Numbers can be expressed as a fraction in a/b where a, b are integers and $b \neq 0$.

Real numbers: The complete set of all rational and irrational numbers. For example, $\sqrt{3}$, -2, $-2/7$ and π are all real numbers.

Roots: A root is a solution to an equation $f(x) = 0$.

Scalar product: A scalar product of two vectors is also called dot product. See **dot product**.

Sequence: A sequence is a series or set of numbers that obeys certain rules. For example, $2, 5, 10, 17, 26, \ldots$ is a sequence as its nth term is $a_n = n^2 + 1$ where $(n = 1, 2, \ldots)$.

Series: The sum of all the terms of a sequence is called a series. For example, $1 + 2 + 3 + \ldots + 99 + 100$ is a series with a formula $1 + 2 + \ldots + n = \frac{n(n+1)}{2}$.

Sine rule: For a triangle with three sides a, b, c and their corresponding angles A, B, C, respectively, the sine rule can be written as $a/\sin A = b/\sin B = c/\sin C$.

Singular: A square matrix $\boldsymbol{A}$ is singular if its determinant is zero. That is $\det(\boldsymbol{A}) = 0$.

Standard deviation: A number which is the square root of the variance of a random variable. See **variance** and **random variable**.

Statistics: The studies of data collection, interpretation, analysis and characterisation of numerical data and sampling.

Stefan-Boltzmann's law: The total energy E per unit surface area per unit time radiated by a black body is given by the Stefan-Boltzmann law $E = \sigma T^4$ where T is the absolute temperature, and $\sigma = 5.67 \times 10^{-8}$ J/s m^2 K^4 is called the Stefan-Boltzmann constant.

Stokes' law: The drag force acted on a spherical object moving in a viscous fluid is $F = 6\pi\mu rv$ where r is the radius of the object and v is its velocity. μ is the viscosity of the fluid.

Surds: A mathematical form. For the nth root, we use $\sqrt[n]{x} = x^{1/n}$. For example, $\sqrt{x}$ denotes the square root of x, $\sqrt[3]{x^2}$ represents the cubic root of x^2.

Taylor's series: A power series to approximate a function in a neighbourhood of a given point $x = a$. Taylor's series can be written as $f(x) = f(a) + f'(a)(x - a) + \frac{f''(a)}{2!}(x - a)^2 + \cdots$.

Transpose: The transpose of a matrix $\boldsymbol{A} = [a_{ij}]$ is obtained by interchanging its rows and columns, i.e., $\boldsymbol{A}^T = [a_{ji}]$. For example, $\boldsymbol{A} = \begin{pmatrix} 2 & -3 \\ 5 & 6 \end{pmatrix}$, we have $\boldsymbol{A}^T = \begin{pmatrix} 2 & 5 \\ -3 & 6 \end{pmatrix}$.

Trigonometrical functions: Functions such as $\sin\theta$, $\cos\theta$ and $\tan\theta$.

Unit vector: A vector with a unit length such as $\boldsymbol{i} = \begin{pmatrix} 1 \\ 0 \end{pmatrix}$.

Variance: A number or a measure of statistical dispersion of a random variable, often denoted by σ^2. The square root of the variance is called the standard deviation σ.

Vector: A vector is a quantity such as force with both a magnitude or length and a direction.

Viscosity: A measure of a fluid's resistance to shear deformation, often denoted by μ with a unit of Pa·s. It is also called dynamic viscosity, Newtonian viscosity or simply viscosity; while kinematic viscosity is defined as $\nu = \mu/\rho$ where ρ is the density of the fluid.

Wave equation: The equation governing the propagation of waves. In the simplest case, the 1-D wave equation is $\frac{\partial^2 u}{\partial t^2} = v^2 \frac{\partial^2 u}{\partial x^2}$ where v is the wave speed.

Young's modulus: A material property that shows how stiff a material is, often denoted by E.

Appendix B

Mathematical Formulae

Quadratic equation

$$ax^2 + bx + c = 0, \qquad x = \frac{-b \pm \sqrt{b^2 - 4ac}}{2a}.$$

Index form

$$a^n = \overbrace{a \times a \times \ldots \times a}^{n}, \quad a^n \times a^m = a^{n+m}, \quad (a^n)^m = a^{n \times m} = a^{nm}.$$

$$a^{1/n} = \sqrt[n]{a}, \quad a^{p/q} = \sqrt[q]{a^p}, \quad (n \neq 0, \ q \neq 0).$$

Binomial theorem

$$(a + b)^n = \binom{n}{0}a^n + \binom{n}{1}a^{n-1}b + \ldots + \binom{n}{k}a^{n-k}b^k + \ldots + \binom{n}{n}b^n.$$

$$\binom{n}{k} = \frac{n!}{k!(n-k)!}, \quad n! = 1 \times 2 \times \ldots \times n, \quad 0! = 1.$$

Series

$$1+2+\ldots+n = \frac{n(n+1)}{2}, \quad a+ar+ar^2+\ldots+ar^{n-1} = \frac{a(1 - r^n)}{1 - r}, \quad (|r| < 1).$$

Exponentials and logarithms

$$\log_b(uv) = \log_b u + \log_b v, \quad \log_b \frac{u}{v} = \log_b u - \log_b v.$$

$$\log_b u^n = n \log_b u, \quad \ln u \equiv \log_e u, \quad \ln e^x = x.$$

Trigonometrical functions

$$\sin^2 \theta + \cos^2 \theta = 1, \quad \cos 2\alpha = \cos^2 \alpha - \sin^2 \alpha = 1 - 2\sin^2 \alpha.$$

$$\sin(\alpha + \beta) = \sin \alpha \cos \beta + \cos \alpha \sin \beta, \quad \cos(\alpha + \beta) = \cos \alpha \cos \beta - \sin \alpha \sin \beta.$$

$$\sin \alpha - \sin \beta = 2 \cos \frac{\alpha + \beta}{2} \sin \frac{\alpha - \beta}{2}, \quad \sin \alpha + \sin \beta = 2 \sin \frac{\alpha + \beta}{2} \cos \frac{\alpha - \beta}{2}.$$

Sine rule

$$\frac{a}{\sin A} = \frac{b}{\sin B} = \frac{c}{\sin C}.$$

Cosine rule

$$a^2 = b^2 + c^2 - 2bc \cos A, \quad c^2 = a^2 + b^2 - 2ab \cos C.$$

Spherical trigonometry

$$S = \frac{\pi R^2}{180°}(A + B + C - 180°), \quad \frac{\sin a}{\sin A} = \frac{\sin b}{\sin B} = \frac{\sin c}{\sin C}.$$

$$\cos \alpha = \cos \beta \cos \gamma + \sin \beta \sin \gamma \cos A.$$

Complex numbers

$$z = a + bi, \ i = \sqrt{-1}, \ |z| = r = \sqrt{a^2 + b^2}.$$

$$z = re^{i\theta}, \ \theta = \arg(z) = \tan^{-1} \frac{b}{a}.$$

$$z^* = a - bi, \ z^* = re^{-i\theta}.$$

Euler's formula

$$e^{i\theta} = \cos \theta + i \sin \theta.$$

$$\sin \theta = \frac{e^{i\theta} - e^{-i\theta}}{2i}, \quad \cos \theta = \frac{e^{i\theta} + e^{-i\theta}}{2}.$$

de Moivre's formula

$$(\cos \theta + i \sin \theta)^n = \cos(n\theta) + i \sin(n\theta).$$

Hyperbolic functions

$$\sinh x = \frac{e^x - e^{-x}}{2}, \quad \cosh x = \frac{e^x + e^{-x}}{2}, \quad \tanh x = \frac{\sinh x}{\cosh x}.$$

$$\cosh(ix) = \cos x, \ \sinh(ix) = i \sin x, \ \cosh^2 x - \sinh^2 x = 1.$$

Differentiation

$$(x^n)' = nx^{n-1}, \ (\sin x)' = \cos x, \ (e^x)' = e^x, \ (\ln x)' = \frac{1}{x} \text{ if } x > 0.$$

$$[\alpha f(x) + \beta g(x)]' = \alpha f'(x) + \beta g'(x).$$

Chain rule

$$\frac{dy}{dx} = \frac{dy}{du} \cdot \frac{du}{dx}.$$

Product rule
$$(uv)' = u'v + uv'.$$

Quotient rule
$$(\frac{u}{v})' = \frac{u'v - uv'}{v^2}.$$

Taylor series

$$e^x = 1 + x + \frac{1}{2!}x^2 + \frac{1}{3!}x^3 + \dots + \frac{1}{n!}x^n.$$

$$\sin x = x - \frac{x^3}{3!} + \frac{x^5}{5!} - \frac{x^7}{7!} + \cdots, \quad \cos x = 1 - \frac{x^2}{2!} + \frac{x^4}{4!} - \cdots.$$

$$f(x) = f(a) + \frac{f'(a)}{1!}(x - a) + \frac{f''(a)}{2!}(x - a)^2 + \dots + \frac{f^{(n)}(a)}{n!}(x - a)^n.$$

Integration

$$\int x^n dx = \frac{x^{n+1}}{n + 1} + C \quad \text{if } n \neq -1, \quad \int \frac{1}{x}dx = \ln x + C.$$

$$\int_a^b [\alpha f(x) + \beta g(x)]dx = \alpha \int_a^b f(x)dx + \beta \int_a^b g(x)dx.$$

Integration by parts

$$\int udv = uv - \int vdu.$$

Fourier series

$$f(t) = \frac{a_0}{2} + \sum_{n=1}^{\infty} [a_n \cos(\frac{n\pi t}{T}) + b_n \sin(\frac{n\pi t}{T})], \quad t \in [-T, T].$$

$$a_0 = \frac{1}{T}\int_{-T}^T f(t)dt, \quad a_n = \frac{1}{T}\int_{-T}^T f(t)\cos(\frac{n\pi t}{T})dt, \quad b_n = \frac{1}{T}\int_{-T}^T f(t)\sin(\frac{n\pi t}{T})dt.$$

Vectors
$\boldsymbol{u} = (u_1, u_2, u_3)^T$, and $\boldsymbol{v} = (v_1, v_2, v_3)^T$,

$$\boldsymbol{u} \cdot \boldsymbol{v} = u_1 v_1 + u_2 v_2 + u_3 v_3.$$

$$\boldsymbol{u} \times \boldsymbol{v} = (u_2 v_3 - u_3 v_2)\boldsymbol{i} + (u_3 v_1 - u_1 v_3)\boldsymbol{j} + (u_1 v_2 - u_2 v_1)\boldsymbol{k}.$$

$$\text{grad}\psi = \boldsymbol{\nabla}\psi = \frac{\partial \psi}{\partial x}\boldsymbol{i} + \frac{\partial \psi}{\partial y}\boldsymbol{j} + \frac{\partial \psi}{\partial z}\boldsymbol{k}, \quad \text{div} \cdot \boldsymbol{u} = \frac{\partial u_1}{\partial x} + \frac{\partial u_2}{\partial y} + \frac{\partial u_3}{\partial z}.$$

$$\Delta\psi \equiv \nabla^2\psi \equiv \boldsymbol{\nabla} \cdot (\boldsymbol{\nabla}\psi) = \frac{\partial^2 \psi}{\partial x^2} + \frac{\partial^2 \psi}{\partial y^2} + \frac{\partial^2 \psi}{\partial z^2}.$$

Matrix

$$\det(\boldsymbol{AB}) = \det(\boldsymbol{A})\det(\boldsymbol{B}), \quad (\boldsymbol{AB})^T = \boldsymbol{B}^T\boldsymbol{A}^T.$$

$$\det(\boldsymbol{A} - \lambda\boldsymbol{I}) = |\boldsymbol{A} - \lambda\boldsymbol{I}| = 0.$$

$$\mathrm{tr}(\boldsymbol{A}) = \sum_{i=1}^{n} a_{ii} = \sum_{i=1}^{n} \lambda_i, \quad \det(\boldsymbol{A}) = \prod_{i=1}^{n} \lambda_i.$$

Error functions

$$\mathrm{erf}(x) = \frac{2}{\sqrt{\pi}} \int_0^x e^{-u^2}\, du, \quad \mathrm{erfc}(x) = 1 - \mathrm{erf}(x), \quad \frac{d\,\mathrm{erf}(x)}{dx} = \frac{2}{\sqrt{\pi}} e^{-x^2}.$$

Gaussian or normal distribution

$$p(x) = \frac{1}{\sigma\sqrt{2\pi}} e^{-\frac{(x-\mu)^2}{2\sigma^2}}, \quad (-\infty < x < \infty).$$

Poisson's distribution

$$p(k) = \frac{\lambda^k e^{-\lambda}}{k!}, \quad (k = 0, 1, 2, 3, ...), \quad \lambda > 0.$$

Sample mean and variance

$$\bar{x} = \frac{1}{n}\sum_{i=1}^{n} x_i = \frac{1}{n}(x_1 + x_2 + ... + x_n), \quad S^2 = \frac{1}{n-1}\sum_{i=1}^{n}(x_i - \bar{x})^2.$$

Propagation of errors

$$\sigma_z^2 = \sigma_x^2\left(\frac{\partial f}{\partial x}\right)^2 + \sigma_y^2\left(\frac{\partial f}{\partial y}\right)^2 + 2\sigma_{xy}\frac{\partial f}{\partial x}\frac{\partial f}{\partial y}, \quad z = f(x, y).$$

Correlation coefficient

$$r = \frac{n\sum_{i=1}^{n} x_i y_i - (\sum_{i=1}^{n} x_i)(\sum_{i=1}^{n} y_i)}{\sqrt{[n\sum_{i=1}^{n} x_i^2 - (\sum_{i=1}^{n} x_i)^2][n\sum_{i=1}^{n} y_i^2 - (\sum_{i=1}^{n} y_i)^2]}}.$$

Newton's method or Newton-Raphson method

$$x_{n+1} = x_n - \frac{f(x_n)}{f'(x_n)}.$$

Trapezium rule

$$I \approx h\left[f_1 + f_2 + ... + f_{n-1} + \frac{(f_0 + f_n)}{2}\right].$$

Simpson's rule

$$I \approx \frac{h}{3}[f_0 + 4(f_1 + f_3 + ... + f_{n-1}) + 2(f_2 + f_4 + ... + f_{n-2}) + f_n].$$

Epilogue

The mathematical techniques we have covered in this book are diverse and some are indeed not easy to follow. This is partly due to the fact that choosing from a wide range of topics itself is a challenging task, and partly due to the multidisciplinary nature of earth sciences, coupled with various levels of complexity which in turn require a wide range of mathematical techniques in the right combinations.

It is true that earth sciences are evolving with ever-increasing emphasis on quantitative modelling. In fact, mathematical modelling and computer simulations have become an integrated part of modern earth sciences. This observation is strengthened by the fact that mathematical and computer modelling will be crucially important to address the *Ten Questions Shaping 21st-Century Earth Sciences* identified by the National Research Council,[1] concerning 1) the formation of Earth and other planets, 2) the missing link during Earth's 'dark age', 3) the origin of life, 4) the Earth's interior, 5) the origin of the plate tectonics and continents, 6) the effect of material properties on large-scale planetary processes, 7) the cause of climate change and the extent of such changes, 8) the interaction of life and Earth, 9) the prediction of earthquakes and volcanic eruptions and their consequences, and 10) the effect of fluid flow and transport on the human environment and ecosystems.

In addition, there are other important issues which I think worth mentioning, and their solutions are equally challenging. Here I list them: 1) efficient use of renewable resources and energy as well as carbon entrapment, 2) long-term change attainability and stability of the earth system, 3) the integrated high-precision Earth observation system, 4) the influence on the earth system induced by the activities of the solar system, especially the Sun, and 5) the long-term impact of all human activities on the planet.

These challenging questions require even more challenging techniques and skills to solve, definitely multidisciplinary and multi-physics as well as multi-scales. After all, we have to be able to model complex systems, to handle massive data, and to deal with uncertainty. Mathematically and computationally speaking, these techniques include nonlinear partial differential equations (PDEs), theory of dynamical systems and complexity, statistical methods such as Monte Carlo simulations, optimisation techniques, modern metaheuristics, and obviously high-performance computing.

Modern approaches to all scientific problems tend to be more computational. Modelling in earth sciences typically uses continuous mod-

---

[1]Committee on Grand Research Questions in the Solid-Earth Sciences, National Research Council, *Origin and Evolution of Earth: Research Questions for a Changing Planet*, (2008).

elling techniques in terms of PDEs. For example, in atmosphere modelling, the mathematical models include many variables with complex 3D geometry and boundary conditions. In addition, we also have to consider chemistry, physics, heat transfer, and fluid dynamics all coupled in an integrated system. So a major trend for such modelling systems will be an integrated approach, because the Earth is after all an integrated system. This will inevitably be coupled with the routine use of high-performance computing.

Recent studies have indicated that many processes such as earthquakes and weather are chaotic under some conditions. Modelling using dynamic systems and complexity analysis will provide some important insight into how such systems behave and to what extent we can influence them by regulating certain parameters.

In addition to the above deterministic approaches to modelling, there are important techniques such as Monte Carlo methods which use a stochastic approach to the solutions of complex problems. Monte Carlo techniques have now been used in almost all areas of modern sciences, from biology to earth sciences. Their advantages are that they can deal naturally and efficiently with higher dimensions with extreme complexity and a large number of degrees of freedom. In addition, data acquisition, analysis and modelling will also be routinely used.

In all these modelling and simulations, we always intend to optimise something, either to minimise the errors, or to maximise the modelling efficiency. In fact, many problems can be recast as optimisation problems. Least-squares method is a good example. In addition, the formulation of finite element method is essentially a minimisation problem. Therefore, many optimisation techniques such as nonlinear programming can be useful when the problems are properly formulated.

Other important emerging techniques such as metaheuristics are becoming powerful. In the last two decades, many new metaheuristic algorithms have emerged, including simulated annealing, particle swarm and other nature-inspired algorithms. These methods will be applied in a wide range of areas, including earth sciences.

Ultimately, we should be able to design an integrated Earth modelling system with an essential set of mathematical models and computer techniques, coupled with real-time high-precision earth observation systems, so that they can capture most of known physics, chemistry, and earth sciences, and more importantly can simulate and predict the performance of systems on earth far into the future. There is no doubt that whatever problems you have to solve, the techniques you will use will be highly mathematical and computational. Perhaps you will become expert at solving some of the challenging problems shaping 21st-century earth sciences.

Index